Introduction to Environmental Technology Systems

The National Skills Academy

ENVIRONMENTAL TECHNOLOGIES

Published in 2011 for the National Skills Academy for Environmental Technologies by:
Nelson Thornes Ltd
Delta Place
27 Bath Road
CHELTENHAM
GL53 7TH
United Kingdom

11 12 13 14 15 / 10 9 8 7 6 5 4 3 2 1

A catalogue record for this book is available from the British Library

ISBN 978 1 908606 00 6

Cover photograph © Kingspan Renewables

Page make-up by Greengate Publishing Services

Text paper contains 100% recycled fibre

Printed and bound in Croatia by Zrinski

Contents

Section one: Introduction

These materials have been developed by the National Skills Academy for Environmental Technologies (NSAET). They are designed to support the delivery of the **QCF Level 3 Award in Understanding the Fundamental Principles and Requirements of Environmental Technology Systems**.

This Award is regulated by the regulatory bodies in England, Wales and Northern Ireland[1] and appears on the Qualifications & Credit Framework (QCF). Learners successfully completing the assessment will achieve 2 QCF credits at Level 3 (see section 1.4 for further information on the QCF).

1.1 Aims and objectives

The aim of this course is to support the development of a fundamental knowledge of micro-renewable energy and water conservation technologies.

On completion of the course the learner should:

- have gained a knowledge of the fundamental working principles, advantages and disadvantages of installation and regulatory requirements for micro-renewable and water conservation technologies
- be prepared, if appropriate, to undertake further units on specialist knowledge and competence to install, commission, handover, inspect, service and maintain micro-renewable energy and water conservation technologies (please note there are pre-requisite entry requirements to undertake these units, and advice should be taken from your Provider)
- be able to successfully complete the assessment to achieve the Level 3 Award in Understanding the Fundamental Principles and Requirements of Environmental Technology Systems.

The learning outcomes of the **QCF Level 3 Award in Understanding the Fundamental Principles and Requirements of Environmental Technology Systems** are that the learner will:

1 know the fundamental working principles of micro-renewable energy and water conservation technologies
2 know the fundamental requirements of building location/building features for the potential to install micro-renewable energy and water conservation systems to exist
3 know the fundamental regulatory requirements relating to micro-renewable energy and water conservation technologies
4 know the typical advantages and disadvantages associated with micro-renewable energy and water conservation technologies.

The course is designed to develop fundamental knowledge only. No part of these materials or the content and assessment of the QCF Award is designed to provide the learner with the occupational competence to install or maintain the technologies covered.

The technologies covered are not exhaustive but include those technologies that are most popular and/or are supported by current or planned Government initiatives or are considered to be key emerging technologies in the short to medium term.

The technologies covered are:

- solar thermal hot water
- heat pumps
- biomass
- solar photovoltaic electricity
- micro-wind power electricity
- micro-hydropower electricity
- micro-combined heat and power (heat led)
- rainwater harvesting
- greywater reuse.

---

1 Office of Qualifications & Examination Regulation (Ofqual) in England; Department for Children, Education, Lifelong Learning & Skills (DCELLS) in Wales; Council for Curriculum Examination and Assessment (CCEA) in Northern Ireland.

Useful weblinks

Ofqual *www.ofqual.gov.uk*
DCELLS *www.wales.gov.uk*
CCEA *www.ccea.org.uk*
QCF *http://www.ofqual.gov.uk/qualification-and-assessment-framework/89-articles/145-explaining-the-qualifications-and-credit-framework*

1.2 The National Skills Academy for Environmental Technologies

The National Skills Academy for Environmental Technologies gives you the green skills you need to profit from the low carbon revolution.

The National Skills Academy for Environmental Technologies was launched in February 2011 and comprises a network of training providers that deliver accredited environmental technologies training. As illustrated in Figure 1.1, the National Skills Academy for Environmental Technologies operates through a central coordinating hub and administrative centre and regional 'lead hub' training providers who work in partnership with other training providers in their area of operation.

Employers and employees using the National Skills Academy for Environmental Technologies will receive the highest quality training, delivered by providers with state-of-the-art facilities and taught by tutors with the most up-to-date knowledge.

These materials are part of a suite of products to support the delivery of first class skills and knowledge in the area of environmental technologies. For more information on further products, please visit *www.nsaet.org.uk*

How do I find out more?

Your first point of contact should be your nearest lead hub provider.

Visit *www.nsaet.org.uk* to identify your nearest lead hub provider.

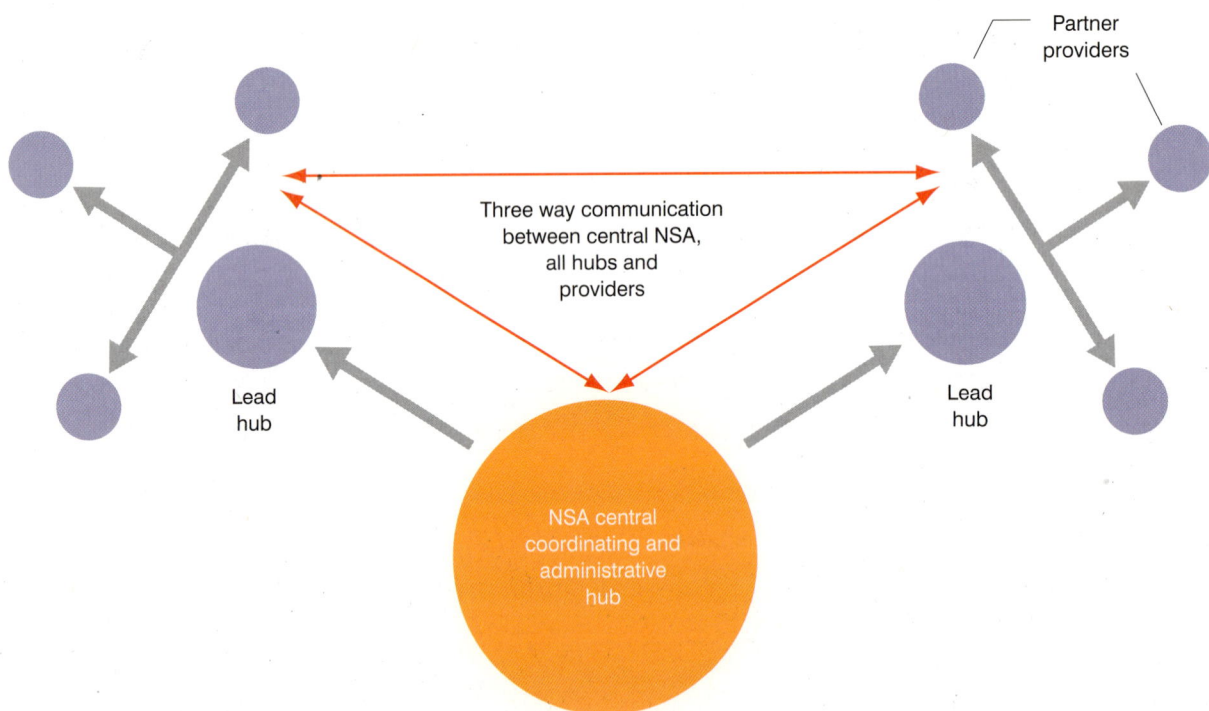

Figure 1.1 Operating model for the National Skills Academy for Environmental Technologies

Academy benefits at a glance

For employers	For employees
✓ Your training will be industry recognised, accredited and linked to the QCF, the national qualification framework which allows your staff's skills to be formally recognised.	✓ You can be safe in the knowledge you have gained industry-recognised accredited qualifications based on National Occupational Standards.
✓ Academy training providers deliver *approved qualifications* using the latest facilities, products, technology and industry knowledge.	✓ You'll be taught by the highest quality trainers in environmental technologies who have been specifically trained to deliver in this area.
✓ All training will be recorded on the Academy Database, accessible by the public, so potential customers can see you are properly qualified to carry out environmental technology work.	✓ Support materials to aid your learning before, during and after the training programmes.
✓ Academy qualifications are fully aligned to the competence requirements for Microgeneration Certification Scheme (MCS) certification – customers must use an MCS certificated installation company to be eligible for Feed-in-Tariff payments and to access other Government incentives.	✓ Your qualification details will be entered onto the central Academy Database which can be accessed by potential customers and/or potential employers to confirm that you have accredited skills in environmental technologies.
✓ Academy qualifications are fully aligned to the competence requirements for Building Regulations Competent Person Scheme membership.	✓ Your training and assessment will be undertaken using the latest environmental technology equipment and products.
	✓ You will have industry-recognised qualifications that will help to future-proof your career and enhance your employment options.

Table 1.1 Academy benefits at a glance

1.3 SummitSkills

The National Skills Academy for Environmental Technologies is a wholly owned, but an independently operated, subsidiary of SummitSkills, the Sector Skills Council for building services engineering. SummitSkills works with employers and other key partners throughout the whole of the UK to ensure that all those who work in the sector are equipped with the right skills, at the right levels, to enable them to be competent, productive, effective and efficient.

SummitSkills also represents the views of the sector's employers, to ensure that building services engineering businesses have a powerful voice at Government level when lobbying for change.

Our **strategic vision** is to have:

'world class skills that drive performance in a profitable and competitive sector'.

To achieve our vision for the sector we will deliver our **mission**:

'to influence and promote investment in skills at all levels'.

The key areas of SummitSkills' work include:

- working with the sector to develop and update National Occupational Standards that set the criteria against which all sector qualifications are developed
- developing and updating Apprenticeship Frameworks, to ensure the sector's apprentices are trained to a high quality

- creating a Sector Qualifications Strategy which details what qualifications are needed now and in the future to help businesses remain properly skilled
- conducting in-depth research and analysis to look at what skills we have in the sector, what we need, and how to bridge any skills gaps
- promoting careers opportunities to young people and careers advisors to help ensure a steady stream of interest in joining the building services engineering sector.

You could have a say in your sector's future…

By taking part in SummitSkills' Strategic Advisory Groups you could have a say on skills issues and give your ideas for what can be done to ensure we have a skilled workforce for the future. This is your opportunity to take part and influence the skills strategies and policies that affect your business.

To discuss how you could take part, contact SummitSkills on 01908 303960 or visit *www.summitskills.org.uk* to find out more about the organisation.

1.4 The Qualifications and Credit Framework

The Qualifications and Credit Framework (QCF) is the national framework for creating and recognising qualifications in England, Wales and Northern Ireland. Its strength lies in the flexibility it gives employers to upskill their workers in the areas needed to keep their businesses competitive. It's also good news for learners who can tailor their learning to support them in their current or future career.

How difficult it is	How long it takes	What it's about
The QCF has 8 levels. Level 1 is the lowest (entry level) up to level 8 (PhD level). Environmental technology qualifications are mainly at QCF Level 3. Level 3 QCF qualifications are similar in level to GCE A Levels.	Each unit within a qualification is given a credit value. One credit = 10 hours of learning for the average learner. Therefore, if a qualification is made up of units whose total credit value is eight, the average time required to complete the qualification would be 80 hours. The number of credits determines the type of qualification: Award (1–12 credits); Certificate (13–36 credits); Diploma (37 credits or more).	A clear title which says what the qualification is about.

Table 1.2 The QCF at a glance

Qualifications on the QCF are made up of units. The units can be combined in a variety of ways to ensure learning is relevant, timely and cost effective. When units are combined into a qualification, its title describes: how difficult it is; how long it takes; and what it's about (see Table 1.2).

Further information on the QCF is available from: *www.ofqual.gov.uk* in England; *www.ccea.org.uk* in Northern Ireland; and *www.wales.gov.uk* (DCELLS) in Wales.

QCF environmental technology qualifications

The current QCF environmental technology qualifications are made up of either one, three or five units as illustrated in Figure 1.2.

The blue units are knowledge units and the green units are practical competence units. The orange arrows show how the units achieved for one qualification can be carried forward and used to achieve other qualifications within the environmental technology qualification suite.

The environmental technology qualifications that are currently available are:

- Level 3 Award in Understanding the Fundamental Principles and Requirements of Environmental Technology Systems Awareness
- Level 3 Award in the Installation of Solar Thermal Hot Water Systems
- Level 3 Award in the Installation of Small Scale Solar Photovoltaic Systems
- Level 3 Award in the Installation of Heat Pump Systems (Non-refrigerant Circuits)
- Level 3 Award in the Installation of Water Harvesting and Reuse Systems
- Level 3 Award in the Installation and Maintenance of Solar Thermal Hot Water Systems
- Level 3 Award in the Installation and Maintenance of Small Scale Solar Photovoltaic Systems
- Level 3 Award in the Installation and Maintenance of Heat Pump Systems (Non-refrigerant Circuits)
- Level 3 Award in the Installation and Maintenance of Water Harvesting and Reuse Systems.

Development of qualifications for the design of the above systems is in progress together with qualifications for micro-combined heat and power, micro-wind and micro-hydropower technologies. Qualifications for the installation and maintenance of biomass fuelled systems are also in development.

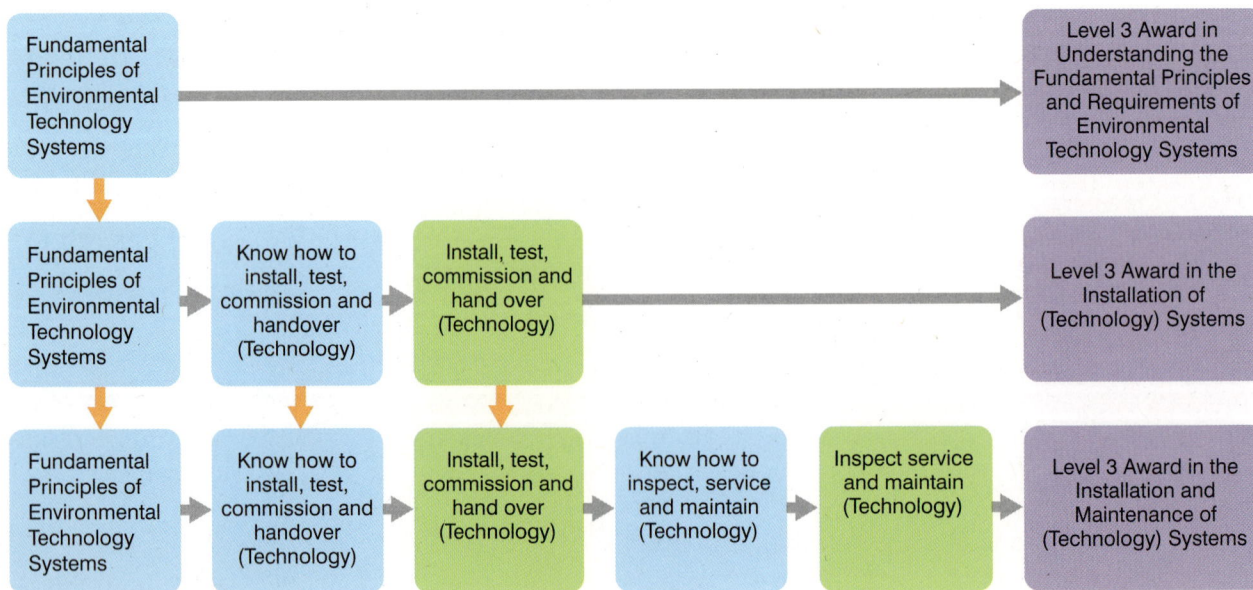

Figure 1.2 QCF environmental technology qualifications

Section two: Background

2.1 Why is renewable energy and sustainability important?

Changes in climate are having an impact across the globe, and their impact can also be seen much closer to home, with ever more extreme weather affecting the UK. Research has shown that carbon dioxide (CO_2) is one of the main greenhouse gases that is causing climate change. In the UK, a large percentage (around 45 per cent) of CO_2 emissions come from everyday activities such as heating the buildings in which we live and work, and air and road travel.

The UK has committed to reduce CO_2 emissions by 34 per cent below 1990 levels by 2020, linked to a target to reduce CO_2 emissions by 80 per cent below 1990 levels by 2050. These targets are very challenging.

Reducing CO_2 emissions, however, is not the only area of concern. Issues such as rapidly reducing fossil fuel supplies and fuel poverty are also driving Government policy to reduce energy demand, improve energy efficiency and increase the use of renewable energy and sustainable technologies.

Consumer interest in renewable and sustainable energy has never been so great, with cutting edge research being undertaken to support new technologies. Investing in such technologies for homes and businesses is becoming increasingly affordable, and this coupled with support from the Government in the form of financial incentives, is seeing a boom in the installation of green technologies.

Useful weblinks:

Department of Energy and Climate Change
www.decc.gov.uk
Act on CO_2 *www.direct.gov.uk/en/ Environmentandgreenerliving/Thewiderenvironment/ index.htm*
Overview of Energy Bill 2010–2011
www.decc.gov.uk/assets/decc/legislation/energybill/427- energy-security-and-green-economy-bill.pdf
Average annual domestic energy bills
www.decc.gov.uk/en/content/cms/statistics/prices/ prices.aspx
Annual report on fuel poverty 2010
www.decc.gov.uk/assets/decc/Statistics/ fuelpoverty/610-annual-fuel-poverty-statistics-2010.pdf

2.2 When is it appropriate to install environmental technologies?

As described in the previous section, environmental technology systems have a key role to play in helping to mitigate climate change and improve sustainability. However, before considering the installation of environmental technology systems it is essential that the first step is to reduce demand followed by improving energy efficiency. Once these steps have been taken it is appropriate to consider the installation of environmental technology systems.

Some examples of steps to reduce energy demand, improve energy efficiency and install environmental technology systems are illustrated in Figure 2.1.

3	Install environmental technologies • Low or zero carbon technologies • Recycling technologies	Solar hot water Solar photovoltaic electricity Heat pumps Water harvesting and recycling
2	Improve efficiency • of energy usage • of water usage	Insulate lofts and pipes Insulate walls (cavity and solid walls) Install double glazed windows Install draught proofing Fit low flow-rate taps/showers
1	Reduce demand • for energy • for water	Switch off lights and appliances Turn heating thermostat down Wash clothes at 30°C Fit a smart meter Energy advice/assessment

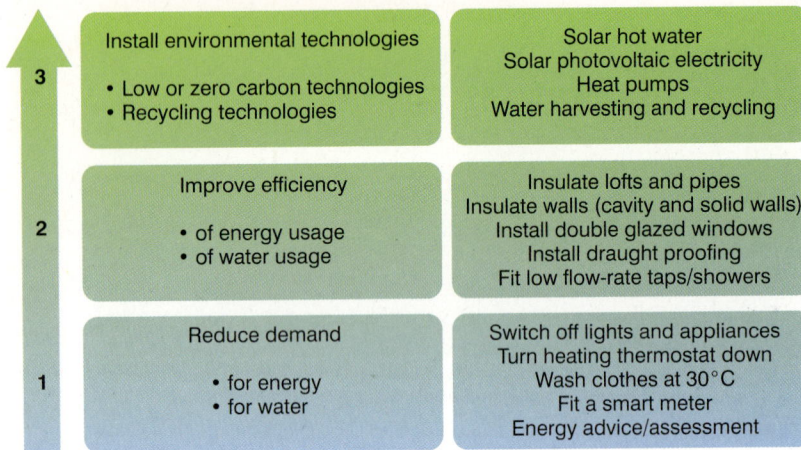

Figure 2.1 Steps to reduce energy demand

Useful weblinks

Energy Savings Trust *www.energysavingstrust.org.uk*
Tips on how to stop wasting energy
   www.energysavingtrust.org.uk/Easy-ways-to-stop-wasting-energy/Stop-wasting-energy-and-cut-your-bills/Tips-to-help-you-stop-wasting-energy
Home improvements and products
   www.energysavingtrust.org.uk/Home-improvements-and-products
How to save energy in the home
   www.direct.gov.uk/en/Environmentandgreenerliving/Thewiderenvironment/index.htm

Section three:
Commercial awareness

3.1 Overview of current and emerging national consumer incentive/financial support schemes

Whilst policy and schemes are subject to change, and the latest information should be sought from the appropriate agency, here is an overview of the current schemes and information on planned initiatives to support the implementation of environmental technologies and promote sustainability.

Feed-in Tariffs

The Feed-in Tariffs (FiTs) scheme is a Government scheme that rewards consumers who generate electricity through technologies, such as:

- hydropower
- windpower
- anaerobic digestion
- solar photovoltaics
- micro-combined heat and power (limited scope).

There are also bonus payments for surplus electricity exported to the electricity supply grid. In order to be eligible for FiTs payments, the products used and the contractor undertaking the installation work must be certificated through the Microgeneration Certification Scheme (see link below).

Renewable Heat Incentive

The Department of Energy and Climate Change (DECC) has released preliminary information on the Renewable Heat Incentive (RHI).

The RHI is the mechanical equivalent of the electrical Feed-in Tariff scheme. Consumers who install an eligible renewable heat technology system will receive financial support in the form of pence per kilowatt hour of renewable heat energy generated. The duration of the support is 20 years for all technologies and the level of support will vary depending on the type and size of technology. Rates in the range of 1.9p to 8p per kilowatt hour have been proposed. The preliminary information states that the RHI will be introduced in two phases:

- Phase 1 – non-residential buildings
- Phase 2 – residential premises.

DECC has indicated that the technologies which will qualify for the RHI are:

- biomass
- solar thermal
- ground source and water source heat pumps
- on site biogas
- deep geothermal
- energy from waste
- injection of biomethane into the gas grid.

Green Deal

The Green Deal is the Coalition Government's flagship policy for improving the energy efficiency of buildings in Great Britain. The Green Deal will create a new financing mechanism to allow a range of improvement measures to reduce energy demand and improve energy efficiency measures to be installed in people's homes and businesses at no upfront cost. The improvement measures to be included under the Green Deal have not yet been finalised but a range of measures have been indicated under the following headings:

- heating, ventilation and air conditioning
- building fabric measures (insulation, windows etc.)
- lighting
- water heating
- microgeneration.

Useful weblinks:

Feed-in Tariffs
www.direct.gov.uk/en/Environmentandgreenerliving/Thewiderenvironment/index.htm
Microgeneration Certification Scheme
www.microgenerationcertification.org/

Renewable Heat Incentive
www.decc.gov.uk/en/content/cms/what_we_do/ uk_supply/energy_mix/renewable/policy/renewable_ heat/incentive/incentive.aspx and *http://www. rhincentive.co.uk/*

Green Deal
www.decc.gov.uk/en/content/cms/what_we_do/ consumers/green_deal/green_deal.aspx

3.2 A profitable future in environmental technologies – make sure your business is ready!

If the UK is to meet its carbon reduction targets, it needs companies and contractors with the skills and competence to design, install and maintain environmental technologies. With the right training, this is where you could benefit.

The FiTs scheme, the RHI and the Green Deal (see section 3.1), in particular, present a huge need for skilled personnel and an even bigger opportunity for business profit.

Your business can benefit…

These schemes, together with future changes to building regulations and the Code for Sustainable Homes, mean that your business will need the right team for the job if it's to win new contracts.

The commercial opportunity presented through the introduction of the FiTs scheme, the RHI and the Green Deal, and the growth of consumer awareness of the need to improve the efficiency of their homes and business premises is immense. However, all schemes and initiatives have a mandatory requirement for the design and installation of the technologies to be undertaken by an accredited installer. Without the required skills and competence there will be no opportunity for a business, be it a sole trader business or a large company, to benefit.

They're added to existing skills, not carried out alone…

Environmental technology skills are typically an extension of skills within existing building services engineering sector occupations such as electrical installation and plumbing and heating. So current sector workers, providing they have the right level of competence, such as NVQ or SVQ (or earlier equivalent), will simply need to undertake short-duration top-up training to upskill and add environmental technologies to their existing skills set.

The opportunities can't be ignored

By considering what skills you have within your business and taking steps to get prepared, you could gain a significant advantage over your competitors as more and more projects specify the need for environmental technologies.

Even at domestic level, homeowners need to be made aware of how to make their home more efficient, not only by maximising their existing equipment but also through the new technologies that are available to them. If your business has the skills and knowledge in this area, you can stay ahead of the game.

What you need to do next

Visit *www.nsaet.org.uk* to find out what environmental technologies are on offer and where you can access training.

Make sure you're ready for the low to zero carbon revolution. Prepare today for business success tomorrow.

Useful weblinks:

Code for sustainable homes
www.communities.gov.uk/planningandbuilding/ sustainability/codesustainablehomes/

3.3 Need help to set up, develop or grow your business?

If you need free and impartial help to set up or develop your business or are seeking support to grow your business in the area of environmental technology, help is at hand.

SummitSkills EnterpriseEssentials is a free portal for anyone in the building services engineering sector who is starting up a business now, already running and growing their business or considering starting a business in the future.

SummitSkills EnterpriseEssentials has been devised and is backed by specialists in the development and nurturing of all types and sizes of business. The skills profiles are based on internationally renowned National Occupational Standards devised and maintained by SFEDI (the Small Firms Enterprise Development Initiative Ltd). All information is checked for accuracy, and all training providers, mentors, courses and other programmes are subject to various quality assurance schemes.

SummitSkills EnterpriseEssentials will even help you find and liaise with a local business adviser when you feel the time is right.

Find out more about SummitSkills EnterpriseEssentials at *www.summitenterprise.co.uk*.

Section four: Heat producing technologies

Approximately 60–70 per cent of household carbon dioxide (CO_2) emissions come from energy usage for space heating and domestic hot water. CO_2 accounts for approximately 85 per cent of all greenhouse gas emissions in the UK. The UK target is, by 2050, to have reduced greenhouse gas emissions by at least 80 per cent below the emission levels that existed in 1990. This target is linked to an interim target of a reduction in emissions of at least 34 per cent below 1990 emission levels by 2020. To achieve this target the use of renewable energy technologies for space heating and domestic hot water will need to increase significantly.

There are a number of renewable energy technologies available which will provide energy to heat space or water. The heat producing technologies included in this section are:

- solar thermal hot water
- heat pumps
- biomass fuelled systems.

These technologies are well developed and are becoming more established in the market with a wide range of installer and consumer choice. The popularity of these technologies is likely to increase as a result of the RHI and the Green Deal.

Solar thermal hot water systems

Introduction

Solar thermal hot water systems have been available in the UK since the 1970s and utilise the free, renewable energy from the sun to heat hot water. The technology is now well developed with a large choice of system types and configurations to suit many different applications.

System categories

There are two main categories of solar thermal hot water system:

1 passive systems
2 active systems.

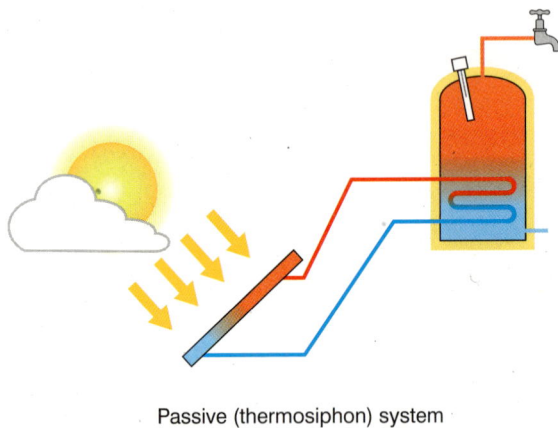

Passive (thermosiphon) system

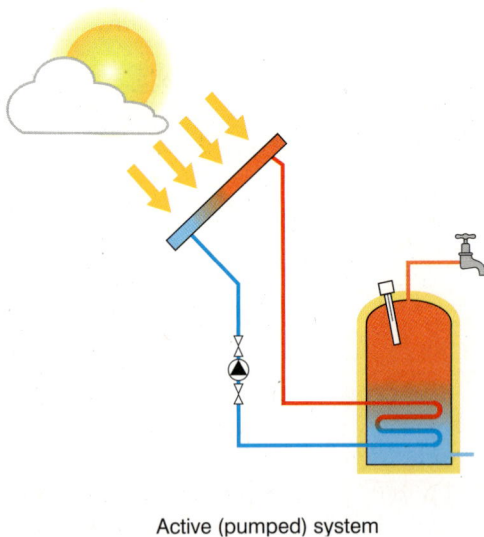

Active (pumped) system

Figure 4.1 Passive and active solar hot water systems

In passive systems, the system circulation takes place by the natural thermosiphon or convection process where heated water expands, becomes lighter and rises, and as water cools it contracts in volume, becomes heavier and falls. For this process to work the solar collector needs to be mounted below the storage cylinder. As this arrangement is not as practical in the UK as it is in warmer countries, the majority of systems installed in the UK are active systems.

In an active system, a circulating pump is included to provide the system circulation. This allows greater flexibility and choice in the positioning of the solar collector(s).

Overview of active solar thermal hot water system components

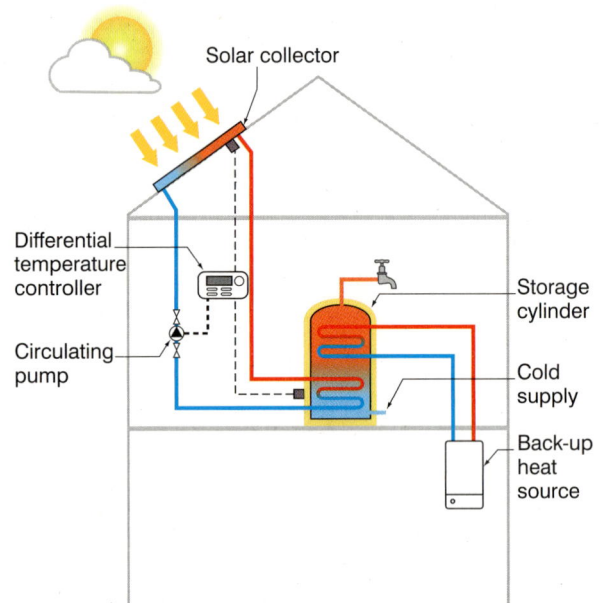

Please note that due to the intended purpose of these materials, some system components are not shown. This is not an installation diagram.

Figure 4.2 Overview of solar thermal hot water systems components (indirect active system)

Solar collector

The function of the solar collector is to absorb the Sun's heat energy and transfer it to the system heat transfer fluid as the heat transfer fluid circulates around the system.

There are two main types of solar collector:

1 flat plate
2 evacuated tube.

Figure 4.3 Example of a flat plate collector installation

Flat plate collectors

Flat plate collectors can be either glazed or unglazed. The intended application will typically determine the selection of glazed or unglazed collectors. Solar thermal systems to heat domestic hot water typically include glazed collectors; solar thermal systems to heat swimming pools typically include unglazed collectors.

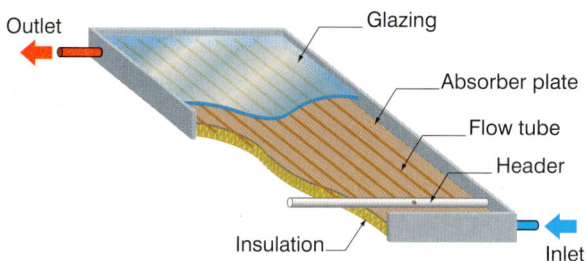

Figure 4.4 Example of a flat plate glazed collector

Evacuated tube collectors

An evacuated tube collector comprises several individual evacuated tubes that are connected to a manifold/header pipe arrangement.

Figure 4.5 Example of an evacuated tube collector installation

The individual evacuated tube collectors come in two main types:

1 heat pipe
2 direct flow.

Heat pipe evacuated tubes contain a sealed copper heat pipe that contains a small volume of a liquid medium that evaporates at reasonably low temperatures. As the liquid medium evaporates, the hot vapour rises and the heat is transferred to the system heat transfer fluid. The cooled vapour then returns to the bottom of the heat pipe to enable the process to repeat itself. To operate, heat pipe evacuated tubes must be installed to a minimum incline as specified by the manufacturer.

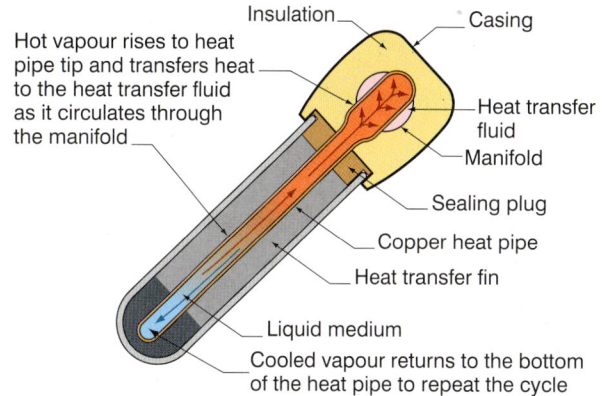

Figure 4.6 Example of a heat pipe evacuated tube

Direct flow evacuated tubes contain a loop of copper pipe that is directly connected to the collector manifold. The system heat transfer fluid circulates around the system and flows directly through the loop of copper pipe within the evacuated tube, enabling direct heat transfer to take place within the evacuated tube itself. As explained above, in the UK, most systems are 'active' with the heat transfer fluid being circulated using a circulating pump. In active systems, the direct flow evacuated tubes can be installed horizontally as well as inclined and vertically. This makes direct flow evacuated tubes suitable for horizontal, inclined and vertical mounting applications.

Differential temperature controller

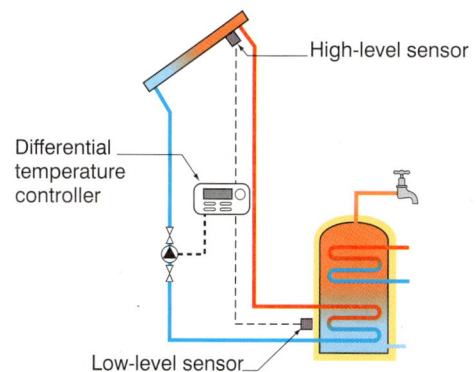

Figure 4.7 Solar thermal hot water system differential temperature controller

The differential temperature controller (DTC) controls the operation of the circulating pump. The DTC is linked to high level and low level temperature sensors. The DTC only allows the system circulating pump to operate when there is:

1 solar energy available; and
2 a demand for water to be heated.

Circulating pump

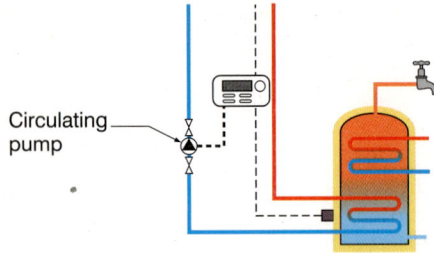

Figure 4.8 Solar thermal hot water system circulating pump

As the name implies, the circulating pump circulates the system heat transfer fluid (either water or glycol depending upon the type of system) around the solar hot water circuit enabling the heat absorbed by the solar collector(s) to be transferred to the stored hot water.

As mentioned above, the operation of the circulating pump is controlled by the differential temperature controller.

Back-up heat source

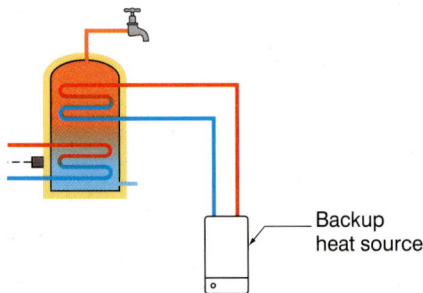

Figure 4.9 Solar thermal hot water system back-up heat source

In the UK, solar thermal hot water systems require a back-up heat source to heat the stored domestic hot water when there is either:

1 insufficient solar energy to heat the water fully; or
2 no solar energy to heat the water.

Where a central heating system is installed, the boiler typically provides the back-up heat source for the solar hot water system. Where no central heating system is installed, the back-up heat source is typically provided by an electric immersion heater located in the solar storage cylinder.

Storage cylinder

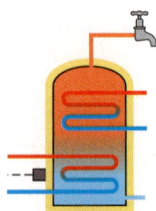

Figure 4.10 Example of a twin coil solar cylinder

The storage cylinder stores the domestic hot water and allows for the heat transfer from the solar collector circuit to the stored domestic hot water. A popular cylinder type is the twin coil cylinder (Figure 4.10). This type of cylinder incorporates a lower solar heating coil and a higher back-up heating coil. Some cylinders may also include a shunt pump to circulate the stored water within the cylinder when just the back-up heat coil is in operation.

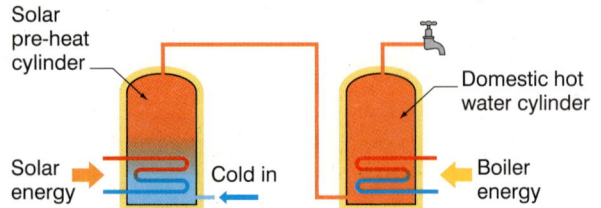

Figure 4.11 Separate solar preheat and storage cylinder arrangement

An alternative arrangement is to use a separate solar preheat cylinder as illustrated in Figure 4.11.

Overview of direct active solar thermal hot water systems

An alternative to an indirect system is the direct solar hot water system. In this type of system, the domestic hot water that is stored in the cylinder is directly circulated through the solar collector.

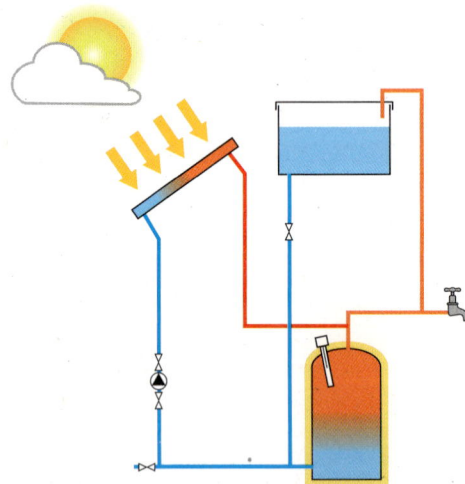

Please note that due to the intended purpose of these materials, some system components are not shown. This is not an installation diagram.

Figure 4.12 Example of a direct solar hot water system layout

This type of system can be installed as a new system or can be added to an existing hot water system. However, it is essential that all system components are compatible with the system design. For example, as the domestic hot water is circulated through the solar collector it is not possible to use a water/glycol mix heat transfer fluid – therefore some components such as the solar collector need to be freeze tolerant.

Regulatory requirements relating to solar thermal hot water systems

The installation of a solar thermal hot water system will require compliance with a number of regulatory requirements including health and safety, water and electrical regulations. A competent installation contractor will have a detailed knowledge of these regulations and will ensure compliance.

Within this section we consider two primary regulatory requirements in relation to solar hot water systems:

1 building regulations
2 planning regulations.

Building regulations

Building regulations set standards for design and construction which apply to most new buildings and many alterations to existing buildings. There are separate building regulations for:

- England and Wales
- Northern Ireland
- Scotland.

The building regulations contain various sections, a number of which have relevance or might have relevance to solar hot water system installation work as summarised in Table 4.1.

Section of regulations	Relevance or potential relevance to solar hot water system installation work
Structural stability	Where solar collectors and other components put load on the structure, in particular the potential effect of wind-uplift loads on the structure
Fire safety	Where holes for pipes, etc. might reduce the fire resistant integrity of the building structure
Site preparation and resistance to moisture	Where holes for pipes, etc. might reduce the water resistant integrity of the building structure
Resistance to the passage of sound	Where holes for pipes, etc. might reduce sound proof integrity of the building structure
Sanitation, hot water safety and water efficiency	Hot water system safety
Conservation of fuel and power	Energy efficiency of the system
Electrical safety in dwellings	Safe installation of electrical controls and components

Table 4.1 Relevance or potential relevance of building regulations to solar hot water system installation work

Town and country planning regulations

As with the building regulations, there are separate town and country planning regulations for:

- England and Wales
- Northern Ireland
- Scotland.

The Town and Country Planning (General Permitted Development) Order 1995 and subsequent amendments, allow local authority planning departments to grant certain 'permitted development' rights. The installation of a solar hot water system will often be classed as permitted development. Table 4.2 summarises the typical solar hot water system permitted development rights for England and Wales.

Building mounted collectors are typically permitted developments providing:	Stand-alone collectors are permitted developments providing:
- the solar collectors are not installed above the ridgeline and do not project more than 200 mm from the roof or wall surface. - the solar collectors are sited, so far as is practicable, to minimise the effect on the appearance of the building - the solar collectors are sited, so far as is practicable, to minimise the effect on the amenity of the area - the property is not a listed building - the property is not in a conservation area or in a World Heritage site.	- the array is no higher than 4 m - the array is sited at least 5 m from boundaries - the size of array is limited to 9 m² or 3 m wide and 3 m deep - the array is not being installed within the boundary of a listed building - in the case of land in a conservation area or in a World Heritage site the array will not be visible from the highway - only one stand-alone solar installation is being installed.

Table 4.2 Summary of the typical solar hot water system permitted development rights for England and Wales

The local planning authority should be consulted for clarification, particularly for installations to flats and non-dwelling building types.

Listed building consent may be required even if planning permission is not required.

Building location and feature requirements for the potential to install a solar thermal hot water system

The following building and location factors will need to be considered:

- a suitable orientation for the mounting of the solar collectors

- a suitable tilt angle for the mounting of the solar collectors
- overshading that might be introduced by adjacent structures or obstructions
- the availability of a suitable solar collector mounting structure
- compatibility with any existing hot water system.

Collector orientation

In the UK, we tend to relate a south facing garden in our homes to the availability of the most sunshine throughout the day. The same applies in relation to solar hot water systems. The ideal collector orientation is south facing. Orientations between south east and south west will also provide good results.

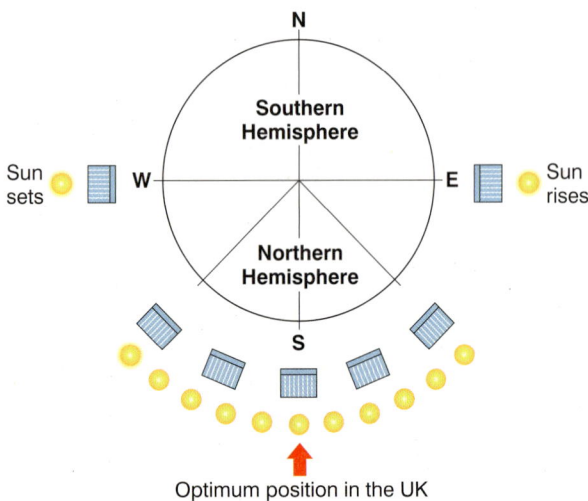

Figure 4.13 Collector orientation options

For buildings with suitable east and west facing roof areas, a split collector system is possible with solar collectors mounted on both east and west facing roof slopes.

Collector tilt

As well as orientation, the 'tilt' of the solar collector is also a key factor that determines the amount of solar energy that is transferred from the sun to the solar hot water system.

Collector tilt is the angle that the solar collector is mounted at from the horizontal plane.

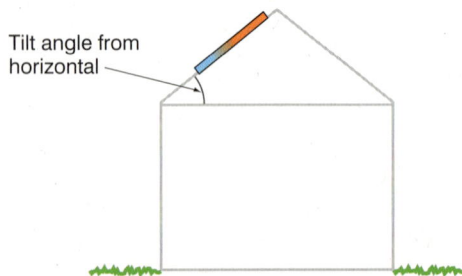

Figure 4.14 Solar collector tilt angle

Where a pitched roof already exists, the tilt is typically determined by the roof pitch. The optimum tilt is approximately 35°. However, a tilt of between 30° and 40° from the horizontal is considered to be close to optimum.

Where there is no pitched roof available, it is possible to mount some types of solar collectors on vertical and horizontal surfaces. Solar collectors may also be mounted on purpose built, roof or ground mounted support frames to provide the required tilt. However, these types of installation may require more design consideration and consultation with product manufacturers, and so on.

Table 4.3 provides examples of annual solar radiation data (kWh/m^2) for various solar collector orientation and tilt arrangements.

Tilt of collector	Orientation of collector				
	South	South east/ south west	East/west	North east/ north west	North
Horizontal	961				
30°	1073	1027	913	785	730
45°	1054	997	854	686	640
60°	989	927	776	597	500
Vertical	746	705	582	440	371

(Source: Table H2, SAP, 2009. Based on a Sheffield, UK location)

Table 4.3 Examples of annual solar radiation data (kWhm^{-2}) for various solar collector orientation and tilt arrangements

Overshading

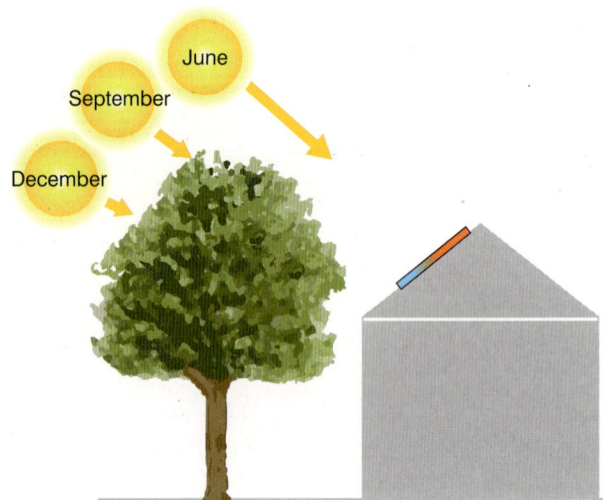

Figure 4.15 Example of an overshading issue

Any overshading of the solar collector(s) will have an impact on how much of the available solar energy is collected and transferred to the solar hot water system. When considering the potential for overshading, it is necessary to consider the position of the sun in the sky at various times of the year. As shown in Figure 4.15, an obstruction might not introduce any overshading when the sun is higher in the sky during the summer months, but might introduce considerable overshading during months when the sun is lower in the sky. The impact of overshading can be up to a 50 per cent reduction in potential system performance. Table 4.4 provides more information on the potential impact of overshading on system performance.

Overshading	Percentage of sky blocked by obstacles	Impact of overshading (Percentage reduction in potential system performance)
Heavy	>80%	50%
Significant	60–80%	35%
Modest	20–60%	20%
None or very little	<20%	none

(Source: Table H4, SAP, 2009)

Table 4.4 Potential impact of overshading on system performance

Collector mounting structure

The collector mounting structure must be suitable in terms of being:

- of sufficient area (m^2)
- able to accommodate the loads imposed by the system
- in good condition.

Typically a minimum of 3–4 m^2 of suitable collector mounting area is needed with approximately 1–1.5 m^2 of area being required per occupant.

Solar collectors will impose some direct weight load onto the roof structure and the design and condition of the structure must be able to accommodate this. However, the wind uplift loads for flat plate collectors will be much higher than the weight loads and the suitability of the structure to cope with wind uplift loads must be taken into account. Solar collector manufacturers will typically provide advice and guidance on the structural requirements for mounting their products.

Installing a solar collector to a roof that is in a poor state of repair is not recommended as replacement or repair of the roof will require the removal of the solar collector. Any roof repairs or refurbishment should be carried out prior to installing the solar collector(s).

Compatibility with existing hot water system

Existing hot water systems come in various types and configurations. Three types of systems are illustrated in Figure 4.16.

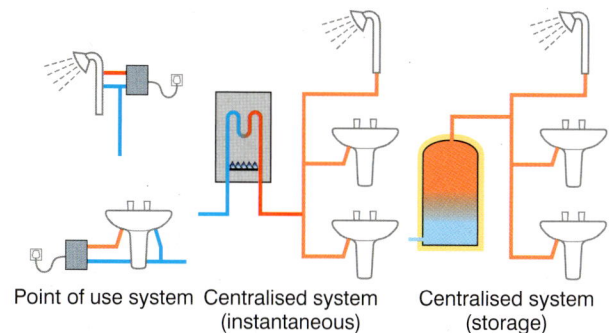

Point of use system   Centralised system (instantaneous)   Centralised system (storage)

Figure 4.16 Types and configurations of existing hot water systems

Point of use systems and instantaneous centralised systems are **not** normally suitable for use with solar hot water systems. Traditionally, centralised storage was the most common type of hot water system installed in the UK. Since the 1980s, however, the number of combination boiler installations has increased significantly.

Combination boilers provide instantaneous centralised hot water. Some combination boilers will accept preheated water but many are designed to take in only cold water. It is also possible to modify some combination boilers so that they operate as a regular boiler which can then be connected to a twin coil solar cylinder. It is essential that the boiler manufacturer is consulted before either connecting a solar thermal hot water system to a combination boiler or modifying a combination boiler to operate as a regular boiler.

Advantages and disadvantages of solar thermal hot water systems

Table 4.5 gives some typical advantages and disadvantages of a solar hot water system.

Advantages	Disadvantages
Reduces carbon dioxide emissions	Not compatible with all existing hot water systems
Solar energy is free and therefore energy costs will be reduced	Less solar energy is available in the winter months
Relatively low maintenance is needed	Initial installation costs can be off-putting
Improves Energy Performance Certificate ratings	Needs a back-up heat source

Table 4.5 Typical advantages and disadvantages of a solar hot water system

Heat pump systems

Introduction

Heat pump systems extract available heat from a natural source such as the ground or air and release it at a higher temperature.

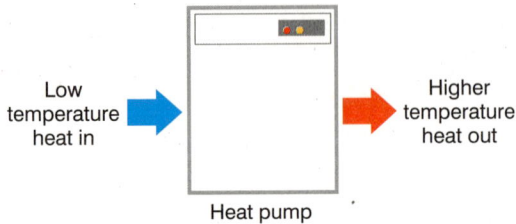

Figure 4.17 Basic principles of a heat pump

Heat pumps can be used for space heating, to heat domestic hot water and to heat swimming pools.

Some heat pumps can also work in reverse and convert high temperature heat to a lower temperature.

How do heat pumps work?

Most heat pumps make use of the mechanical vapour compression cycle commonly known as the refrigeration cycle to convert heat from one temperature to another.

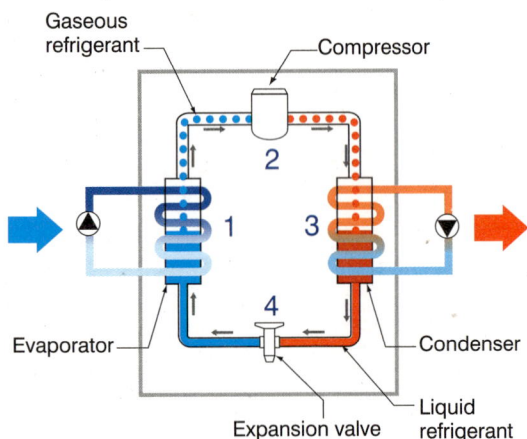

Figure 4.18 Example of a mechanical vapour compression cycle heat pump

1 The low temperature heat (heat source) enters the evaporator which is a heat exchanger. A refrigerant within the evaporator is at a cooler temperature than the heat source and heat is transferred from the source into the refrigerant causing the refrigerant to boil and evaporate.
2 The refrigerant, now in a gaseous state, enters the compressor resulting in a rise in the temperature and pressure of the refrigerant.

3 The refrigerant continues its course through the condenser (which is also a heat exchanger) transferring the higher temperature heat to either an air or water distribution circuit (often referred to as the 'heat sink' or emitter circuit).
4 The refrigerant, now in a liquid state again and at a cooler temperature, enters the expansion valve, which reduces its pressure and temperature to its initial state at the evaporator. The cycle then repeats itself.

Energy input and output ratios and the coefficient of performance

Heat pumps are classified as a 'low' carbon technology because they need some electrical energy to operate.

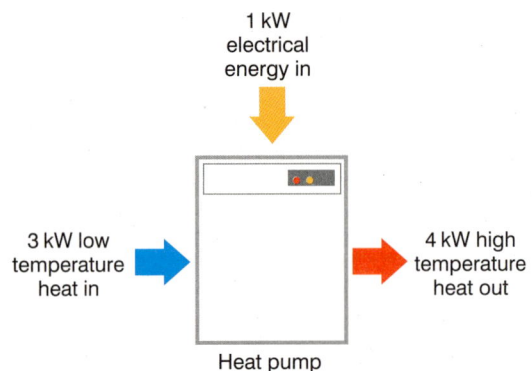

Figure 4.19 Example of heat pump energy input and output

Depending on the application, operating conditions and type of heat pump utilised, heat pump energy output can be 300–500 per cent more than the electrical energy input.

Heat pump efficiency is referred to as the coefficient of performance (COP). In its simplest form COP relates to **heating output** divided by the **electrical power input**. For this example, the COP is 4.0, calculated as follows:

Heating output (4 kW) ÷ Electrical power input (1 kW) = 4.0

Types of heat pump system

Heat pump technology is able to utilize low temperature heat from an air, ground or water source to produce higher temperature heat for use in ducted air or piped water heat sink systems. The type of heat pump unit must be selected in relation to the intended heat source and heat sink arrangement.

Exhaust air heat pumps that receive exhaust heat from other systems are also available but are not covered in this course.

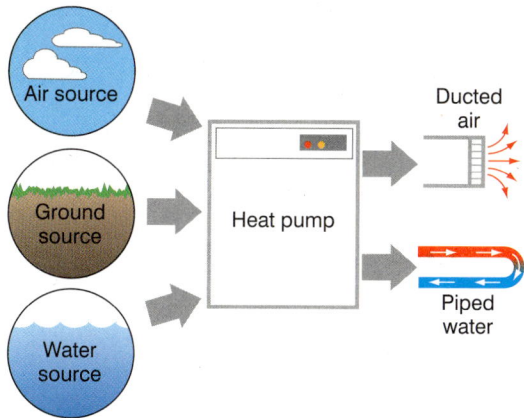

Figure 4.20 Typical heat pump input and output options

Air source heat pumps

A variety of heat pump system arrangements are available using the external air as the heat source. The options include:

- internal and external 'packaged' heat pump units
- split heat pump units
- air or water heat sink circuits.

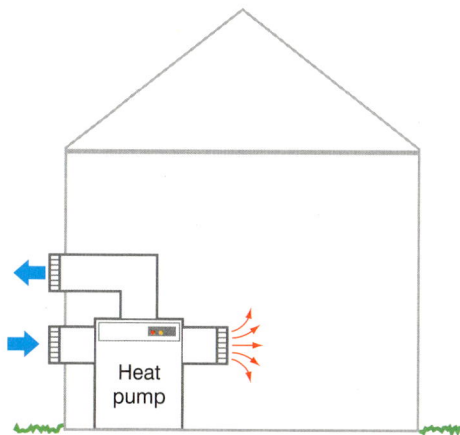

Figure 4.21 Example of an air source packaged heat pump unit connected to a ducted air heat sink circuit

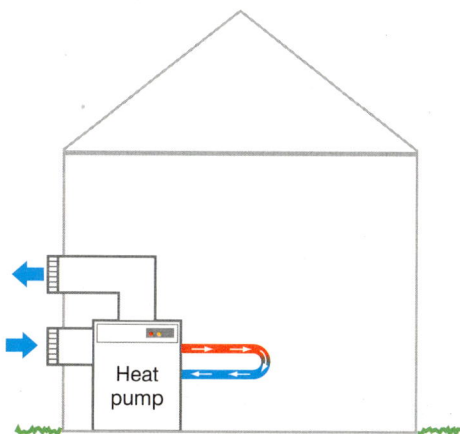

Figure 4.22 Example of an air source packaged heat pump unit connected to a piped water heat sink circuit

Figures 4.21 and 4.22 show internal 'packaged' unit air source heat pumps that receive and reject air through ducts that pass through the external wall of the building. External 'packaged' units are also available.

Split heat pump units are also available. Split heat pump units include an external fan coil (evaporator) unit that is linked by refrigerant circuit pipework to an internal unit that contains the rest of the heat pump components. Figures 4.23 and 4.24 illustrate typical split air source heat pump arrangements.

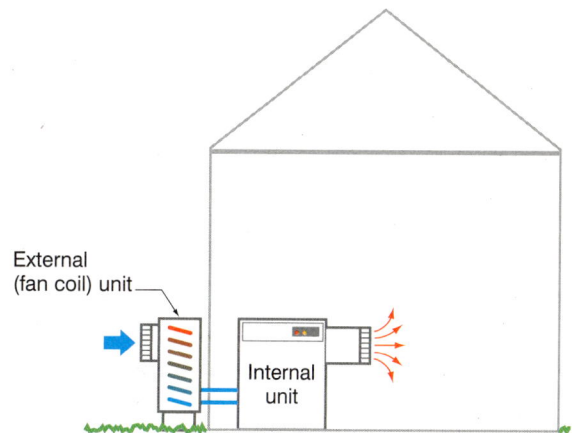

Figure 4.23 Example of a split air source heat pump unit connected to a ducted air heat sink circuit

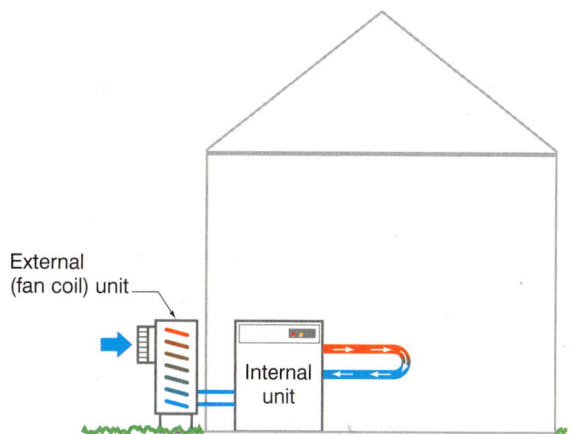

Figure 4.24 Example of an air source to water heat sink with a split heat pump unit

Some fan coil units can be noisy and the noise level that will be generated needs to be considered at the design stage.

Air source heat pumps can and do operate effectively at sub-zero ambient air temperatures. However, the efficiency of all heat pumps, including air source heat pumps, improves when a higher source temperature is available.

Ground source heat pumps

A variety of heat pump system arrangements are available using the ground as the heat source. The options include:

- horizontal or vertical ground collector circuits
- internal and external 'packaged' heat pump units
- air or water heat sink circuits.

Some example arrangements are illustrated in Figures 4.25–4.28.

'Slinky' type collectors are a type of compact collector that can be laid horizontally or vertically in ground trenches. Slinky type collectors are sometimes used where the available ground area (m^2) is limited. The heat pump manufacturer should be consulted before any compact type collector is included in the system design.

Vertical borehole collector loops may be used where the geothermal conditions support the use of a ground source heat pump but where the available ground

Figure 4.27 Example of a ground source packaged heat pump unit connected to a vertical borehole closed loop ground collector circuit and a piped water heat sink circuit

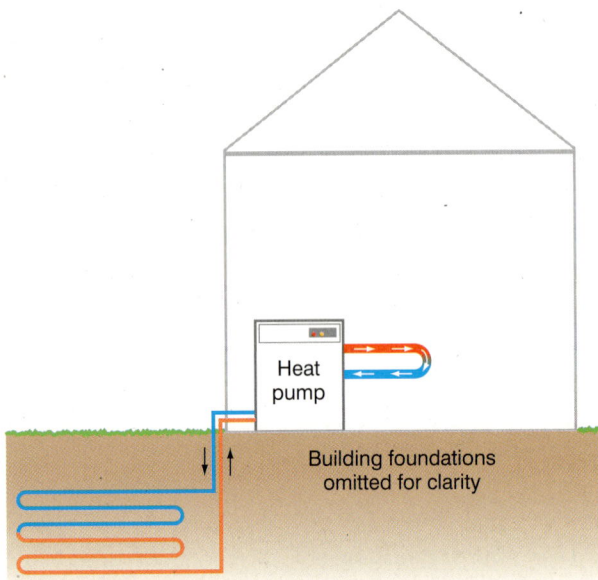

Heat pump

Bore hole

Building foundations omitted for clarity

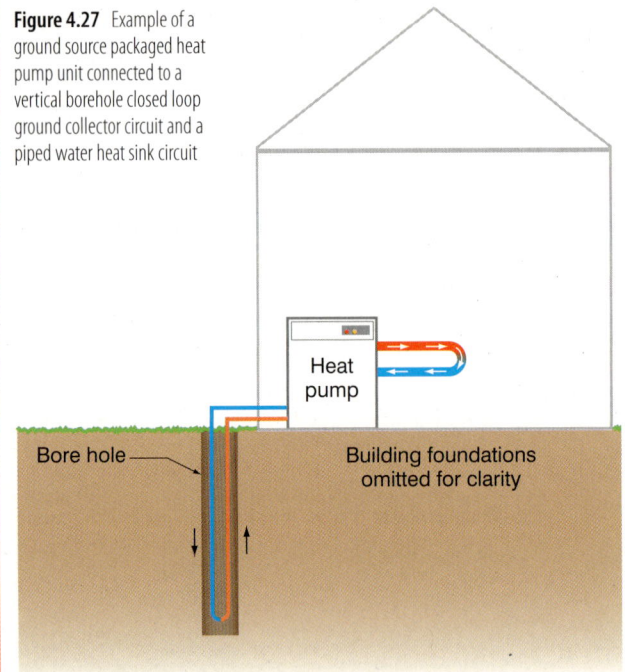

Heat pump

Building foundations omitted for clarity

Figure 4.25 Example of a ground source packaged heat pump unit connected to a horizontal closed loop ground collector circuit and a piped water heat sink circuit

Figure 4.28 Example of a ground source packaged heat pump unit connected to a vertical borehole open loop ground collector circuit and a piped water heat sink circuit

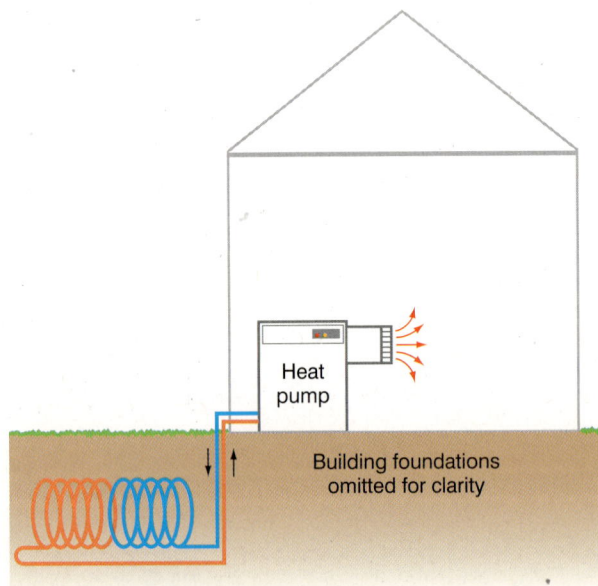

Heat pump

Building foundations omitted for clarity

Figure 4.26 Example of a ground source packaged heat pump unit connected to a closed loop 'Slinky' type ground collector circuit and a ducted air heat sink circuit

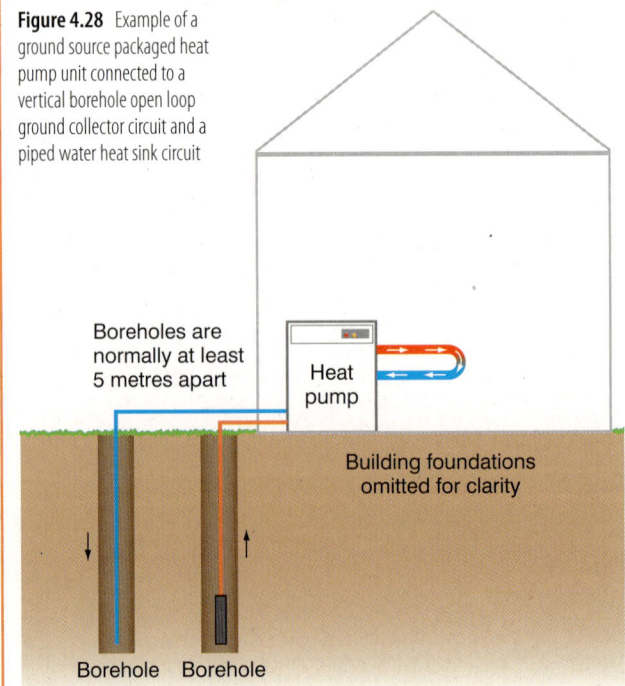

Boreholes are normally at least 5 metres apart

Heat pump

Building foundations omitted for clarity

Borehole   Borehole

area (m²) is limited. This type of installation requires a specialist drilling rig to be used to create the borehole. A specialist contractor is normally used to undertake the drilling operation.

A vertical borehole 'open' loop arrangement includes two boreholes. The collector circuit fluid flows naturally through the open collector circuit from the open-ended return pipe to the open-ended flow pipe. This type of arrangement requires the availability of a suitable geothermal water source and an Environmental Agency Abstraction Licence.

Water source heat pumps

Where a suitable water source exists such as a lake or a pond, this can be a very effective alternative to a ground source collector circuit.

For illustration purposes the collector in Figure 4.29 is shown in a vertical position, but water source collectors are simply laid on the bottom of the lake or a pond and weighted as necessary to keep them in place. Open loop water source collector circuits (not illustrated) are also an option.

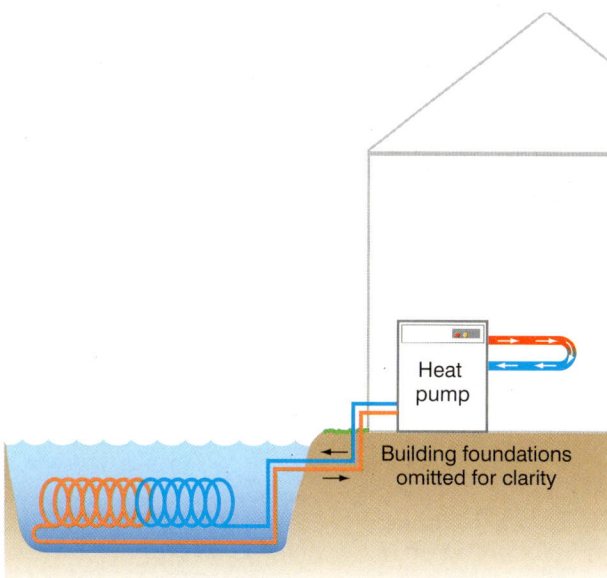

Figure 4.29 Example of a water source packaged heat pump unit connected to a closed loop Slinky type collector circuit and a piped water heat sink circuit

Heat pump system piped water 'heat sink' circuits

Piped water heating systems that incorporate a modern condensing type boiler have a flow temperature of 80°C and a return temperature of 60°C providing a mean water temperature (MWT) of approximately 70°C. Heat pump systems will typically have a flow temperature of between 35°C and 55°C and a return

temperature of between 25°C and 45°C providing a MWT of between 30°C and 40°C.

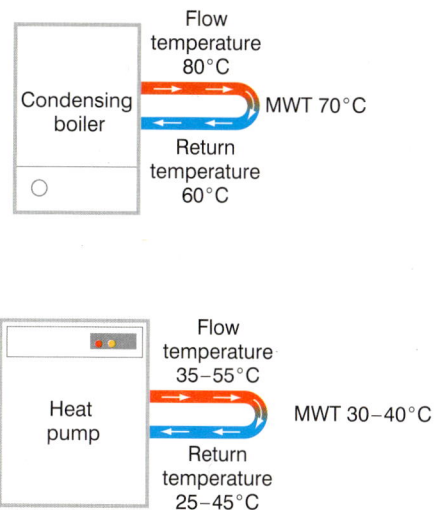

Figure 4.30 System mean water temperature comparison

The lower mean water temperature dictates that some types of heat emitters and hot water storage cylinders are more suitable for use with heat pump systems than others.

Domestic hot water storage

Heat pumps can be used to heat a domestic hot water storage cylinder. Standard type indirect hot water storage cylinders are not typically suitable for use with heat pump systems due to the lower mean water temperature and the limited surface area of the heat transfer coil within the cylinder.

The 'tank-in-tank' design of a hot water cylinder provides a large surface-to-surface contact between the inner tank containing the stored domestic hot water and the outer tank containing the heating circuit hot water.

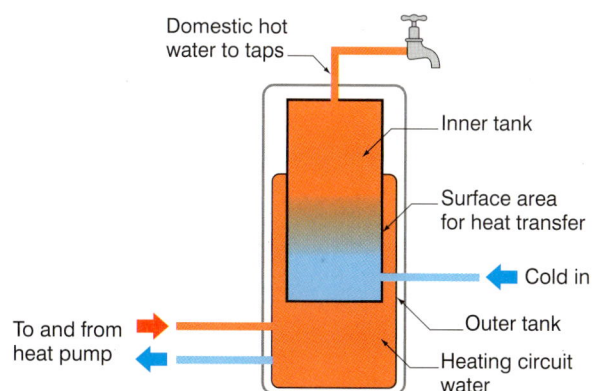

Figure 4.31 Example of a tank-in-tank hot water cylinder

This design is very suitable for a heat pump system. Some heat pump units include an integrated tank-in-tank cylinder.

For most heat pump installations that provide heat for domestic hot water, it is necessary to include a 'boost' or auxiliary heat source to raise the stored water temperature to the standard 60°C domestic hot water storage temperature.

Underfloor heating

Underfloor heating systems operate at a lower MWT than a heating system with radiators. Therefore, underfloor heating is very suitable for use with heat pumps.

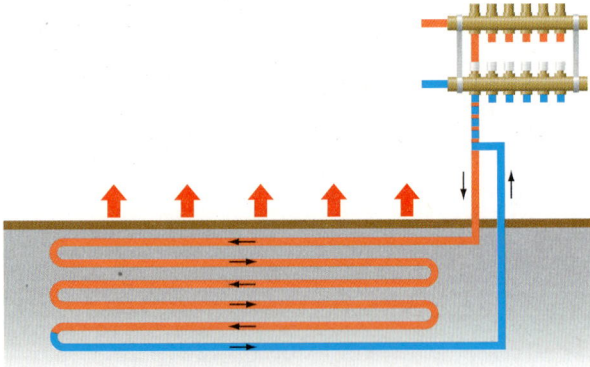

Figure 4.32 Example of an underfloor heating circuit

Panel radiators

Standard type panel radiators can be used with heat pump systems; however, standard type panel radiators are designed to work at a MWT of approximately 70°C.

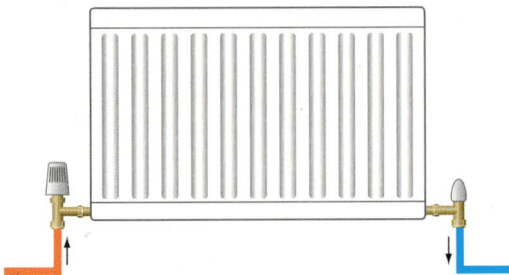

Figure 4.33 Example of a standard panel radiator

Due to the lower MWT in a heating circuit connected to a heat pump, standard panel radiators will need to be oversized to achieve the heat output necessary to meet the design room temperature. The need to oversize standard panel radiators by 100 per cent or more when compared to the size of a radiator required for a system with a MWT of 70°C is not uncommon. The need to oversize by this much can make standard panel radiators a less attractive option for use with heat pump installations. The oversizing requirement increases with lower MWT and reduces with higher MWT. The higher the MWT, however, the lower the coefficient of performance achieved by the heat pump.

When a heat pump is being considered for connection to an existing heating system that includes standard panel radiators, it is necessary to estimate the heat output that will be achieved from each existing radiator at the intended system MWT. It is not uncommon for panel radiators in an existing heating system to have been oversized in relation to design room temperature and building heat loss. Where this is the case it can be an advantage. If existing radiators are oversized and all practicable measures are taken to improve the fabric energy efficiency of the building, the potential to connect a heat pump to an existing heating system will increase.

In addition to standard panel radiators, low temperature high efficiency panel radiators are also available. This type of panel radiator is typically more suitable for a new heat pump system installation where panel radiators are the required type of heat emitter.

Convector heaters

Natural and fanned convector heaters include finned heat exchangers that increase heat output. This design makes natural and fanned convector heaters an option for use with heat pumps.

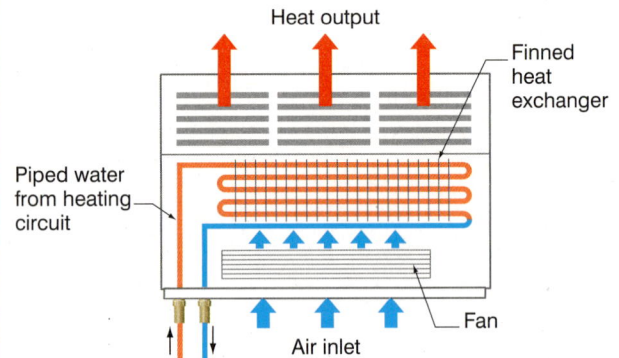

Figure 4.34 Example of a fanned convector heater

As with panel radiators, where natural and/or fanned convector heaters are used, significant oversizing might need to be applied unless the system MWT is increased.

Buffer tanks

Heat pumps are not designed or sized to meet short-term heat loads. For efficient operation a heat pump needs to be able to start up and run for a period of time. Stop-start operation can also shorten the life of the heat pump compressor. The inclusion of a buffer tank that accumulates and stores heated water is a method of meeting short-term heat loads without stop-start operation of the heat pump.

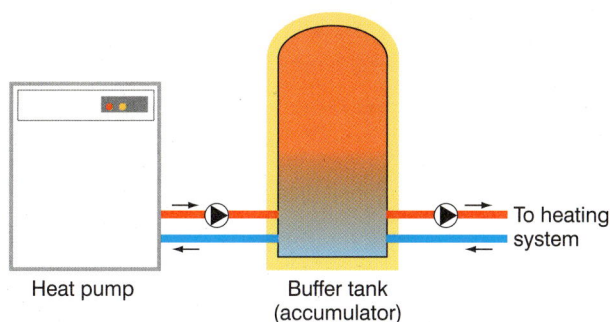

Figure 4.35 Example of a buffer tank installation

Most air source heat pumps, particularly those with an external fan coil unit, need to defrost regularly. Buffer tanks are also useful to provide heat to defrost the fan coil unit. During the defrost cycle the heat pump refrigeration cycle operates in reverse.

Buffer tanks are also useful where an auxillary heat source such as a boiler is being used with a heat pump. A system that includes an auxillary heat source system is known as a bivalent system. A system where the heat pump is the only heat source is a known as a monovalent system.

Regulatory requirements relating to heat pump systems

The installation of a heat pump system will require compliance with:

1 building regulations
2 town and country planning regulations.

Building regulations

Building regulations set standards for design and construction which apply to most new buildings and many alterations to existing buildings. There are separate building regulations for:

- England and Wales
- Northern Ireland
- Scotland.

The building regulations contain various sections, a number of which have relevance or might have relevance to heat pump system installation work as summarised in Table 4.6.

Section of regulations	Relevance or potential relevance to heat pump system installation work
Structural stability	Load on structure might require additional strengthening work
Fire safety	Holes for pipework might affect fire resistant integrity of building
Site preparation and resistance to moisture	Holes for pipes might reduce the water resistant integrity of the building structure
Resistance to passage of sound	Holes for pipes might reduce soundproof integrity
Sanitation, hot water safety and water efficiency	Hot water system safety
Conservation of fuel and power	Energy efficiency
Electrical safety	Safe installation of electrical controls and components

Table 4.6 Relevance or potential relevance of building regulations to heat pump installation work

Town and country planning regulations

As with the building regulations, there are separate town and country planning regulations for:

- England and Wales
- Northern Ireland
- Scotland.

The Town and Country Planning (General Permitted Development) Order 1995 and subsequent amendments, allow local authority planning departments to grant certain 'permitted development' rights.

Installing a ground source or water source heat pump system does not usually need planning permission and should fall within permitted development rights.

Due to potential noise issues, most air source heat pump installations currently require planning permission. However, this is currently being reviewed and as soon as relevant standards and safeguards to deal with noise have been established air source heat pumps are likely to be classified as permitted development.

The local planning authority should be consulted for clarification, particularly for installations in conservation areas and installations to non-dwelling building types.

Listed building consent might be required even if planning permission is not required.

Building location and feature requirements for the potential to install a ground source heat pump system

The following building and location factors will need to be considered:

- the fabric energy efficiency of the building – high fabric energy efficiency is particularly important when considering a heat pump
- adequate space for collector circuit trenches or boreholes
- ground conditions (type of rock strata, etc.)
- access requirements to enable trench excavation or borehole drilling
- the compatibility of any existing heating and hot water systems with the proposed heat pump installation
- a suitable electrical supply
- if auxillary or boost heat is needed and, if so, how it will be provided.

Building location and feature requirements for the potential to install an air source heat pump system

The following building and location factors will need to be considered:

- the fabric energy efficiency of the building – high fabric energy efficiency is particularly important when considering a heat pump
- the options to position the fan coil unit or air intake in a position that has the best ambient air temperature (i.e. ideally not north facing or on an elevation that is exposed to strong winds)
- the level of noise that will be generated by the fan coil unit – is this acceptable to building occupants and neighbouring properties?
- the compatibility of any existing heating and hot water systems with the proposed heat pump installation
- a suitable electrical supply
- if auxillary or boost heat is needed and, if so, how it will be provided.

Building location and feature requirements for the potential to install a water source heat pump system

The following building and location factors will need to be considered:

- the fabric energy efficiency of the building – high fabric energy efficiency is particularly important when considering a heat pump
- adequate water source (lake, pond, etc.) for the collector circuit
- the compatibility of any existing heating and hot water systems with the proposed heat pump installation
- a suitable electrical supply
- if auxillary or boost heat is needed and, if so, how it will be provided.

Advantages and disadvantages of heat pump systems

Table 4.7 gives some typical advantages and disadvantages of a heat pump system.

Advantages	Disadvantages
Reduces carbon dioxide emissions	Might not be suitable for connection to existing heating systems that include panel radiators
Efficiencies between 300–500% are typical	Initial installation costs can be off-putting
Relatively low maintenance is needed	Air source installations can present a noise issue
Improves Energy Performance Certificate ratings	Ground source installations require a large ground area or a borehole

Table 4.7 Typical advantages and disadvantages of a heat pump system

Biomass fuelled systems

Introduction

Biomass is the term given to solid biomass fuels derived from trees. Biomass fuelled systems are generally considered to be carbon neutral. This is because the carbon dioxide released when combustion takes place is equal to the carbon dioxide that was used during the tree growing process. This is illustrated in the biomass cycle shown in Figure 4.36.

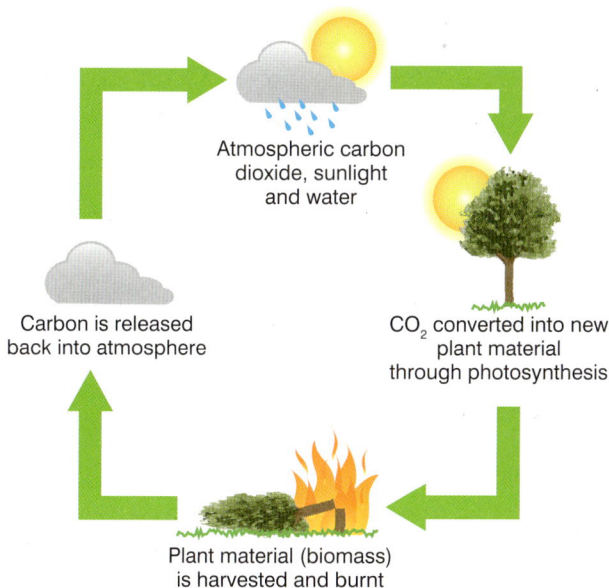

Figure 4.36 The biomass cycle

Biomass fuels

Within this section we focus on only woody biomass fuels. Table 4.8 shows the three main types of biomass fuels.

Logs		Logs have been used to provide heating for thousands of years and are the original biomass fuel. Logs for biomass appliances need to be of maximum length and diameter
Wood chip		Wood chip is typically produced from the small round wood that is left over when trees are felled and logs are harvested but can also be produced from reclaimed timber
Pellets		Wood pellets are made from fine wood particles such as sawdust. They are cylindrical in shape, typically 6 or 8 mm wide (diameter), and 15–30mm long.

Note: Biomass fuels must be stored in a dry environment to minimise the fuel moisture content level. Logs and wood chip also require a ventilated storage area.

Table 4.8 Woody biomass fuels

Types of biomass appliances and systems

There are two main categories of biomass appliance:

1 biomass stoves
2 biomass boilers.

Each type of appliance has a range of fuel type and output options.

Biomass stoves

Pellet burning biomass stoves and log burning biomass stoves are available.

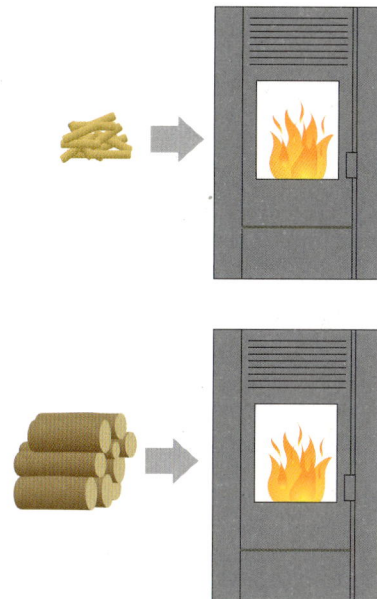

Figure 4.37 Biomass stove fuel options

Pellet burning biomass stoves include an integrated hopper and an auger feed mechanism that transfers the pellets from the hopper to the burner when heat is needed. Log burning stoves require manual loading.

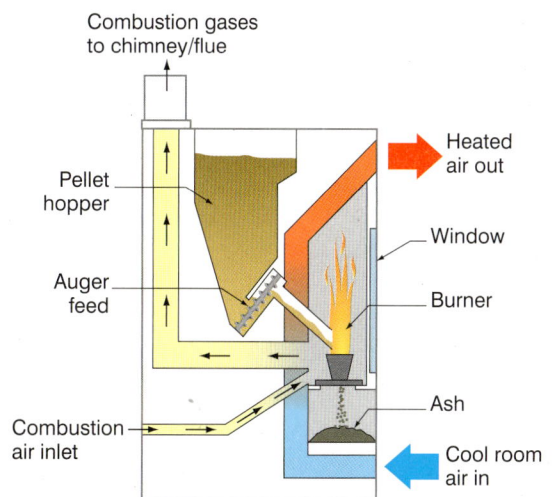

Figure 4.38 Example of a pellet burning biomass stove

The typical heat capacity range for biomass stoves is 5–15 kW but some stoves can be regulated to outputs as low as 2 kW.

Biomass stoves provide a number of heat output options as illustrated in Figure 4.39.

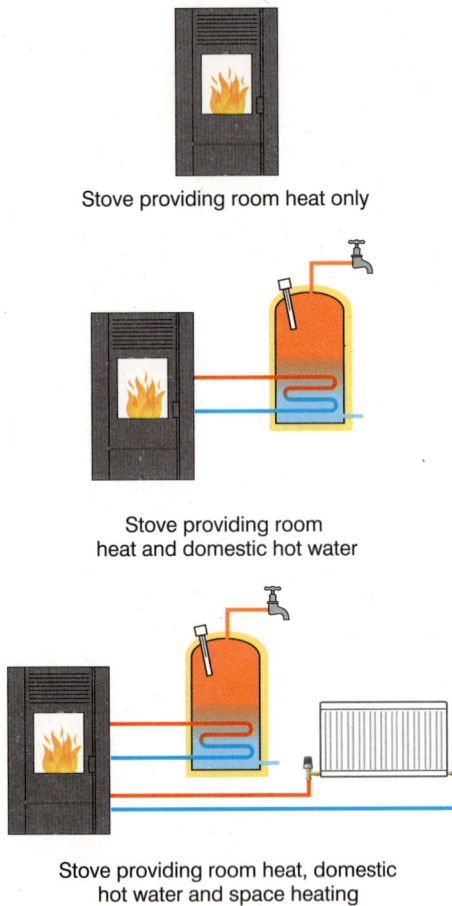

Stove providing room heat only

Stove providing room heat and domestic hot water

Stove providing room heat, domestic hot water and space heating

Figure 4.39 Biomass stove heat output options

Biomass boilers

Biomass boilers are available for all three main types of biomass fuel. Some biomass boilers are multi-fuel boilers.

Figure 4.40 Biomass boiler fuel options

Biomass boilers provide space heating and domestic heat output options as illustrated in Figure 4.41.

The typical minimum heat output rating for biomass boilers is 8 kW for pellet boilers, 12 kW for log boilers and 25 kW for wood chip boilers. Biomass boilers are typically more suited to larger domestic properties, non-domestic applications and communal heating schemes. For smaller domestic properties, a biomass stove that can provide heat for domestic hot water and space heating purposes is often used.

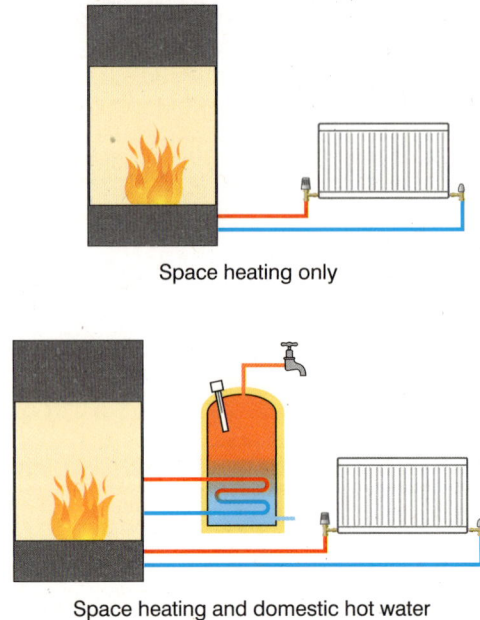

Space heating only

Space heating and domestic hot water

Figure 4.41 Biomass boiler heat output options

Biomass fuel storage and transfer

For smaller pellet burning biomass appliances, pellets are available in sealed bags that can be carried and loaded directly into the appliance. Some biomass pellet burning boilers have attached external fuel hoppers. Although these hoppers need to be filled manually, once filled, the supply will last for several days before the hopper needs to be refilled.

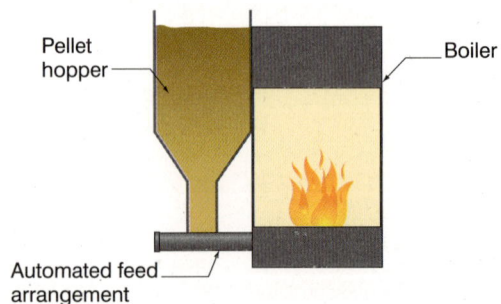

Figure 4.42 Example of a biomass boiler with attached external fuel hopper

Larger pellet burning and wood chip burning biomass boiler installations include an automated fuel transfer arrangement. The biomass fuel is stored in a suitably constructed internal or external, above or below ground,

Supply pipe for pellet deliveries (blown pellet filling)

Silo providing bulk pellet storage

Overhead supply pipe to supply pellets from silo to hopper

Pellet hopper

Boiler

Automated feed arrangement

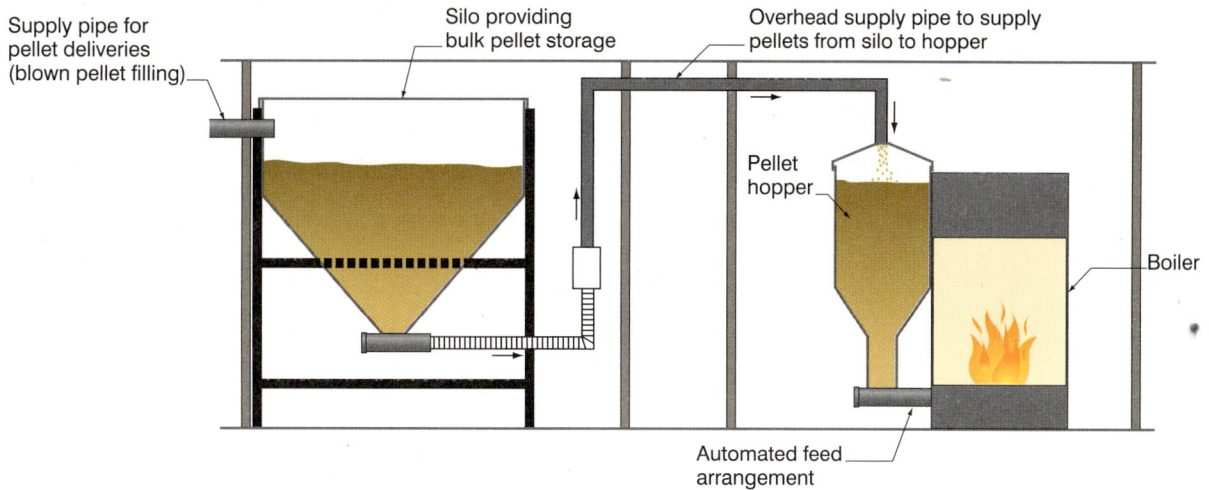

Figure 4.43 Example of a pellet silo providing bulk storage with overhead supply to boiler hopper

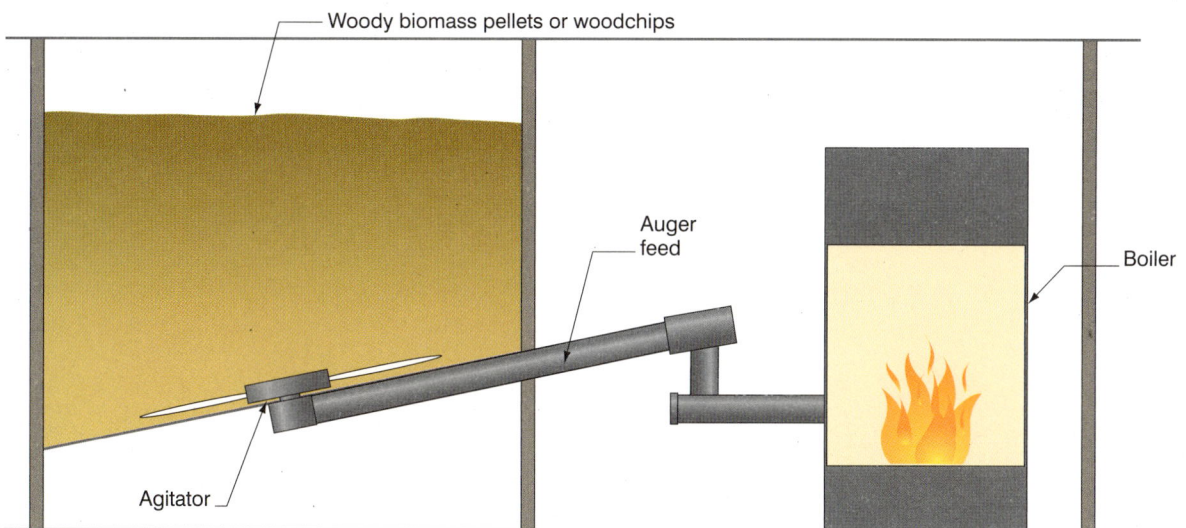

Woody biomass pellets or woodchips

Auger feed

Boiler

Agitator

Figure 4.44 Example of bulk storage with auger feed direct to boiler

bulk storage area and is automatically transferred to the boiler using either an auger feed or overhead transfer arrangement. Pellet deliveries are normally blown into the bulk store; wood chip deliveries are normally tipped or loaded manually into the bulk store.

Figures 4.44 and 4.45 illustrate only two examples of bulk storage and automated fuel transfer options. Many other options exist.

Bulk stores for pellets must be designed to prevent moisture being absorbed by the fuel. Bulk stores for wood chips should be well ventilated to let the wood chips breathe and prevent mould.

The size of the fuel store depends on many factors:

- anticipated fuel requirements
- fuel type
- reliability of deliveries
- space available
- delivery vehicle capacity, etc.

It is normally cheaper to have large loads of fuel delivered providing suitable storage is available.

Log burning biomass boilers need to be loaded manually, but some boilers can accommodate enough logs to burn for three or four days without the need to reload.

Regulatory requirements related to biomass fuelled systems

The installation of a biomass fuelled system will require compliance with:

1 building regulations
2 town and country planning regulations.

25

Building regulations

Building regulations set standards for design and construction which apply to most new buildings and many alterations to existing buildings. There are separate building regulations for:

- England and Wales
- Northern Ireland
- Scotland.

The Building regulations contain various sections a number of which have relevance or might have relevance to biomass fuelled system installation work as summarised in Table 4.9.

Section of regulations	Relevance or potential relevance to biomass fuelled system installation work
Structural stability	Load on structure might require additional strengthening work
Fire safety	Where holes for pipes, etc. might reduce the fire resistant integrity of the building structure
Site preparation and resistance to moisture	Where holes for pipes, etc. might reduce the moisture resistant integrity of the building structure
Resistance to the passage of sound	Where holes for pipes, etc. might reduce the sound proof integrity of the building structure
Sanitation, hot water safety and water efficiency	Hot water safety and water efficiency
Combustion appliances and fuel storage system	Biomass appliances are heat-producing combustion appliances and must be installed safely
Conservation of fuel and power	Energy efficiency of the system and the building
Electrical safety in dwellings	Safe installation of electrical controls and components

Table 4.9 Relevance or potential relevance of building regulations to biomass fuelled system installation work

Town and country planning regulations

As with the building regulations, there are separate town and country planning regulations for:

- England and Wales
- Northern Ireland
- Scotland.

The Town and Country Planning (General Permitted Development) Order 1995 and subsequent amendments, allow local authority planning departments to grant certain 'permitted development' rights.

Planning permission is not normally needed when installing a biomass fuelled system in a house if the work is all internal. However, if the installation requires an outside flue it will normally be permitted development if the flue is on the rear or side elevation of the building and projects to no more than one metre above the highest part of the roof.

If the building is listed or in a designated area, even if the building has permitted development rights, it is advisable to check with the local planning authority before a flue is fitted. Consent is also likely to be needed for internal alterations.

In a conservation area or in a World Heritage site the flue should not be fitted on the principal or side elevation if it would be visible from a highway.

If the project also requires an outside building to store fuel or related equipment the same rules apply to that building as for other extensions and garden outbuildings.

Smoke control legislation and areas

Under the Clean Air Act, local authorities may declare the whole or part of the district of the authority to be a smoke control area. It is an offence to emit smoke from a chimney of a building, from a furnace or from any fixed boiler if located in a designated smoke control area unless it is an **exempted appliance** using an **authorised fuel**.

The Secretary of State for Environment, Food and Rural Affairs has powers under the Act to authorise smokeless fuels or exempt appliances for use in smoke control areas in England. The Department for Environment, Food and Rural Affairs (DEFRA) provides information regarding exempted appliances and exempted fuels.

Local authorities are responsible for enforcing the legislation in smoke control areas and the local authority should be contacted for details of any smoke control areas in their area. The local authority might also be able to provide information regarding exempted appliances and exempted fuels.

Building location and feature requirements for the potential to install a biomass fuelled system

The following building and location factors will need to be considered:

- a suitable flue or chimney system or the potential to install a suitable flue or chimney system must exist

- the flue system must be constructed of, or lined with, a material that is suitable to receive the products of combustion from a biomass appliance
- prefabricated gas appliance flue systems are not suitable for biomass appliances
- the flue system must be fitted with an appropriate terminal to disperse the products of combustion
- a suitable location and arrangement for fuel storage. Factors such as space, moisture, access for fuel deliveries and the frequency of fuel deliveries must be considered.

Advantages and disadvantages of biomass fuelled systems

Table 4.10 gives some typical advantages and disadvantages of a biomass fuelled system.

Advantages	Disadvantages
Biomass is a carbon neutral technology	Requires a suitable flue/chimney system
Does not rely on building orientation or weather conditions to operate effectively	Initial installation costs can be off putting
Biomass is generally considered to be an inexhaustible fuel source	Larger appliances typically require a large space to bulk store fuel
Producing biomass fuel is very cheap compared to the cost of finding and extracting fossil fuels	Sometimes considered less suitable for smaller properties

Table 4.10 Typical advantages and disadvantages of a biomass fuelled system

Section five: Electricity producing technologies

The majority of electricity for domestic and commercial use is produced through processes that involve the burning of fossil fuels leading to carbon dioxide emissions. As the UK population grows and technologies such as electric cars evolve, consumer demand for electricity increases. This increasing demand together with rising energy costs and the need to reduce the UK's carbon emissions, means increasing the amount of electricity that is produced from renewable energy sources is a priority for the UK government and for many consumers.

The electricity producing technologies included in this section are:

- solar photovoltaic systems
- micro-wind turbine systems
- micro-hydropower systems.

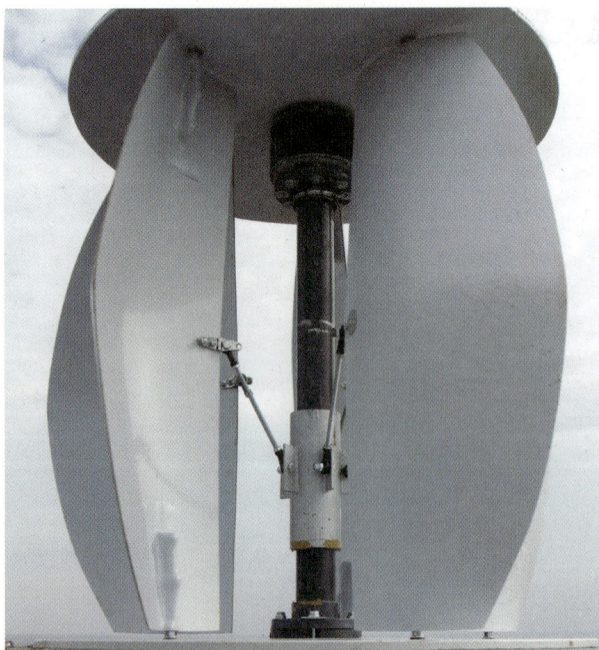

Solar photovoltaic systems

Introduction

The term photovoltaic means the production of electric current at the junction of two substances exposed to light. In basic terms a solar photovoltaic system is a system that uses solar cells to convert light energy from the sun into electricity.

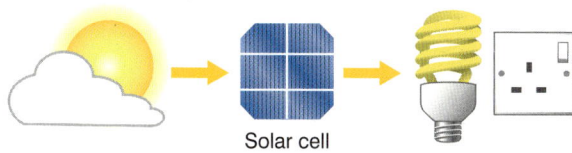

Figure 5.1 Basic principles of solar photovoltaic technology

There are three stages to the working of a solar cell:

1 photons in sunlight hit the solar cell and are absorbed by semiconducting materials within the cell, such as silicon
2 the negatively charged electrons within the solar cell are knocked loose from their atoms, allowing them to flow through the material to produce electricity. The composition of solar cells allows the electrons to move in a single direction only
3 direct current (d.c.) electricity is generated.

A group of solar cells is needed to generate a usable amount of electricity. A group of solar cells is known as a solar photovoltaic module (Figure 5.2).

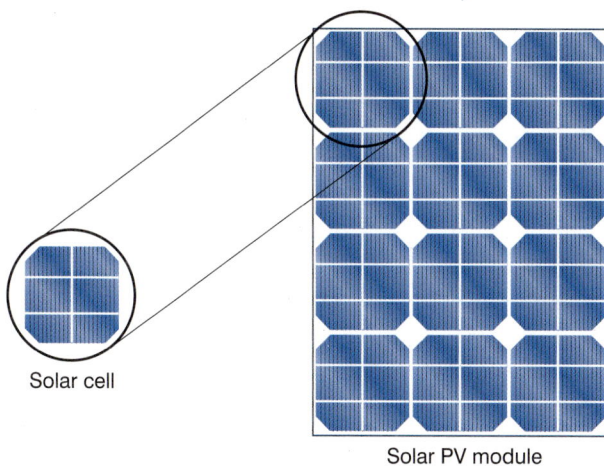

Figure 5.2 A solar photovoltaic module

Solar photovoltaic electricity generation, usually referred to as solar PV, is a zero carbon technology.

Basic system categories

Although there are a number of system types, variations and configurations, solar PV systems fall into two basic system categories:

1 on-grid systems
2 off-grid systems.

On-grid systems

On-grid systems allow any surplus electricity that is generated to be exported to the electricity distribution grid.

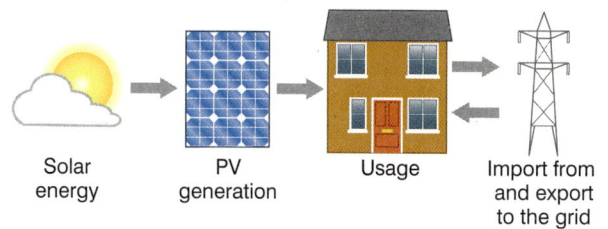

Figure 5.3 Basic principles of an on-grid solar PV system

Off-grid systems

Off-grid systems use a battery bank arrangement to store the electrical power generated for use when needed.

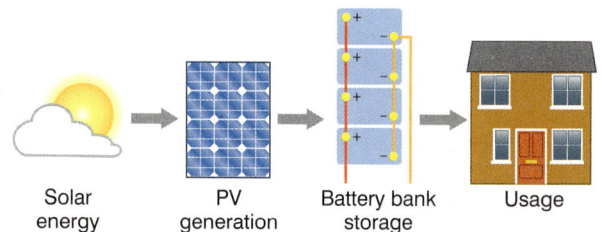

Figure 5.4 Basic principles of an off-grid solar PV system

Some solar PV systems combine on- and off-grid arrangements.

The popularity of on-grid systems has increased significantly since the introduction of the Feed-in Tariff scheme.

Overview of system components

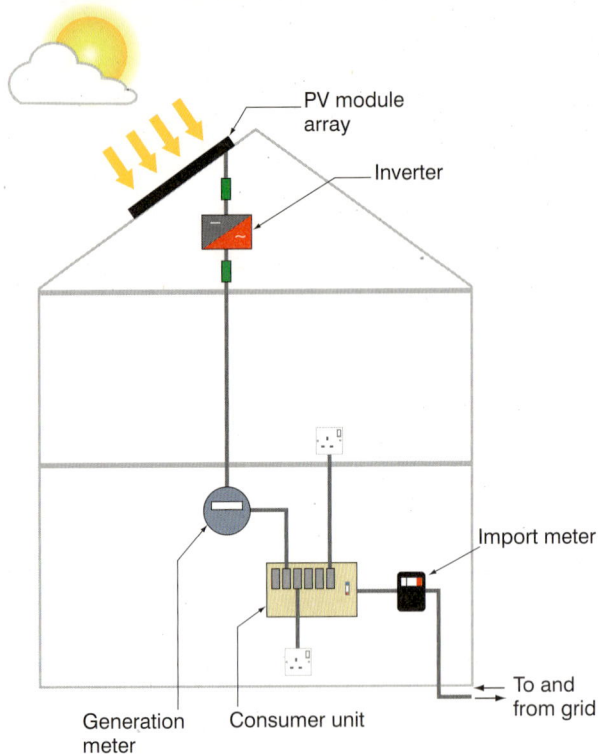

Please note that due to the intended purpose of these materials, some system components are not shown. This is not an installation diagram.

Figure 5.5 Solar PV system component overview (on-grid system)

PV module array

As explained above, a group of solar cells is known as a solar PV module. When installed on its own, a solar PV module is also known as a solar PV array. However, the term array is typically used to refer to a group of solar PV modules.

Solar PV modules come in different forms and types. Table 5.1 lists three common types, together with details of the composition and typical efficiency for each type.

PV module type	Composition	Typical efficiency
Monocrystalline	Thin slices from a single crystal of silicone	15%
Polycrystalline	Thin slices from a block of silicone crystals	12%
Thin film	Thin layer of semiconducting atoms on glass or metal base	7%

Table 5.1 Composition and typical efficiency for different PV module types

Solar PV modules may be mounted above roof surfaces, integrated into roof surfaces, integrated into building facades, and fixed on building mounted or ground mounted support frames. Solar PV roof tiles are also available.

Figure 5.6 Example of a 15-module roof mounted solar PV array

Figure 5.7 Example of an 18-module frame mounted solar PV array

Inverter

Some electrical appliances operate using d.c. electricity, but the most common type of electricity used in our homes and places of work is alternating current (a.c.).

Before d.c. electricity that is generated by a solar PV system can be used with a.c. systems and appliances, the electricity has to be converted from d.c. to a.c. electricity.

It is only possible to export a.c. electricity to the electricity distribution grid. Therefore on-grid a.c. solar PV systems are the most popular and common type of system.

The inverter is the system component that converts electricity from d.c. to a.c. Depending upon the solar PV module layout and size, the d.c. voltages that enter the inverter can be much higher than the 230V a.c. system voltage that is used in the UK.

The inverter can be mounted in the roof space adjacent to the solar PV module location or in the building.

Figure 5.8 Example of an inverter

Consumer unit

The consumer unit or fuse board, as it is sometimes referred to, is the connection point for the solar PV system installation.

Where the existing consumer unit is of a modern type, and has a spare connection circuit point available, it can often be utilized. Older type consumer units will need to be replaced at the time that the solar PV system is installed.

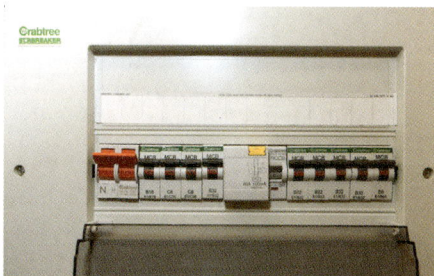

Figure 5.9 Example of a consumer unit

Generation meter

A generation meter is fitted to record how much solar PV electricity has been generated and how much has been exported to the supply grid.

Some energy supplier's incoming supply (import) meters have the capability to perform this function but a generation meter is typically included as part of the solar PV installation.

The generation meter must be of an approved type and located in a position where it is easily accessible for purposes of reading. In the UK, the Office of the Gas and Electricity Markets (OFGEM) is responsible for approving generation meters.

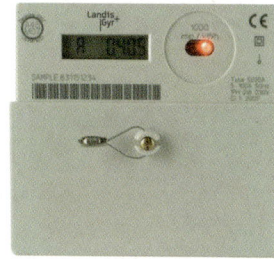

Figure 5.10 Example of a generation meter

Regulatory requirements relating to solar PV systems

The installation of a solar PV system will require compliance with a number of regulatory requirements, including health and safety and electrical regulations. A competent installation contractor will have a detailed knowledge of these regulations and will ensure compliance.

Within this section we consider two primary regulatory requirements in relation to solar PV systems:

1 building regulations
2 planning regulations.

Building regulations

Building regulations set standards for design and construction which apply to most new buildings and many alterations to existing buildings. There are separate building regulations for:

- England and Wales
- Northern Ireland
- Scotland.

The building regulations contain various sections, a number of which have relevance or might have relevance to solar PV system installation work as summarised in Table 5.2:

Section of regulations	Relevance or potential relevance to solar PV system installation work
Structure	Where solar PV modules and other components put load on the structure, in particular wind uplift loads
Fire safety	Where holes for cables, etc. might reduce the fire resistant integrity of the building structure
Site preparation and resistance to moisture	Where holes for cables, etc. might reduce the moisture resistant integrity of the building structure
Resistance to the passage of sound	Where holes for cables, etc. might reduce the sound proof integrity of the building structure
Electrical safety in dwellings	Safe installation of electrical controls and components

Table 5.2 Relevance or potential relevance of building regulations to solar PV system installation work

Town and country planning regulations

As with the building regulations, there are separate town and country planning regulations for:

- England and Wales
- Northern Ireland
- Scotland.

The Town and Country Planning (General Permitted Development) Order 1995 and subsequent amendments, allow local authority planning departments to grant certain 'permitted development' rights.

The installation of building mounted solar PV arrays is typically classed as permitted development, providing:

- the solar modules are not installed above the ridgeline and do not project more than 200 mm from the roof or wall surface
- the solar modules are sited, so far as is practicable, to minimise the effect on the appearance of the building
- the solar modules are sited, so far as is practicable, to minimise the effect on the amenity of the area
- the property is not a listed building
- the property is not in a conservation area or in a World Heritage site.

The local planning authority should be consulted for clarification, particularly for installations to flats and non-dwelling building types.

Listed building consent might be required even if planning permission is not required.

The installation of stand-alone solar PV arrays is typically classed as permitted development providing:

- the array is no higher than 4 m
- the array is sited at least 5 m from boundaries
- the size of the array is limited to 9 m² or 3 m wide and 3 m deep
- the array is not being installed within the boundary of a listed building
- in the case of land in a conservation area or in a World Heritage site, the array will not be visible from the highway
- only one stand-alone solar installation is being installed.

Building location and feature requirements for the potential to install a solar PV system

The following building and location factors will need to be considered:

- orientation of the solar PV array
- tilt of the solar PV array
- adjacent structures or obstructions that introduce overshading
- the availability of a suitable solar PV array mounting structure.

Orientation of the solar PV array

As with solar hot water system collectors, the ideal orientation to mount solar PV arrays is south facing.

Orientations between south east and south west will also provide good results.

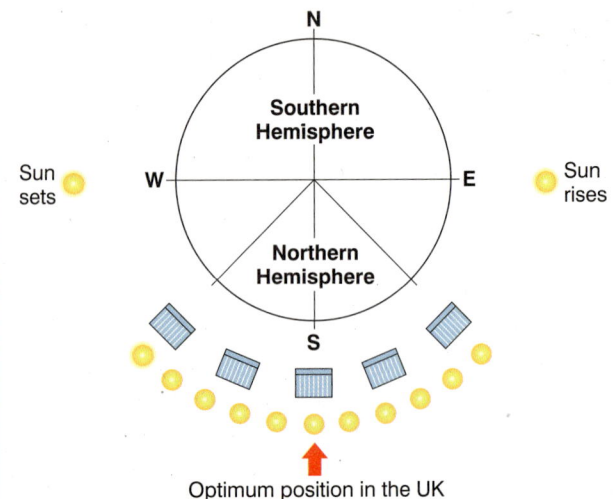

Figure 5.11 Solar PV array orientation

Tilt of the solar photovoltaic array

As well as orientation, the tilt of the solar PV array is also a key factor that determines the amount of solar energy that is harnessed from the sun and converted to electrical energy.

Tilt is the angle that the solar PV array is mounted from the horizontal plane.

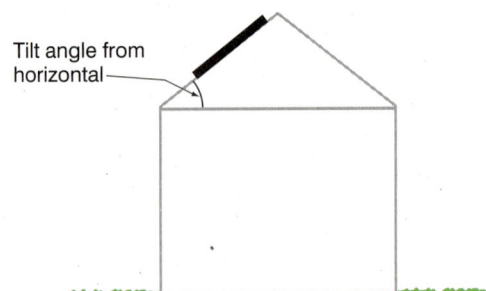

Figure 5.12 Solar PV array tilt

Typically, a tilt of between 30° and 40° from horizontal is considered to be close to optimum.

Adjacent structures or obstructions that introduce overshading

Any overshading of the solar PV array will have an impact on how much solar energy is harnessed from the sun and converted to electrical energy.

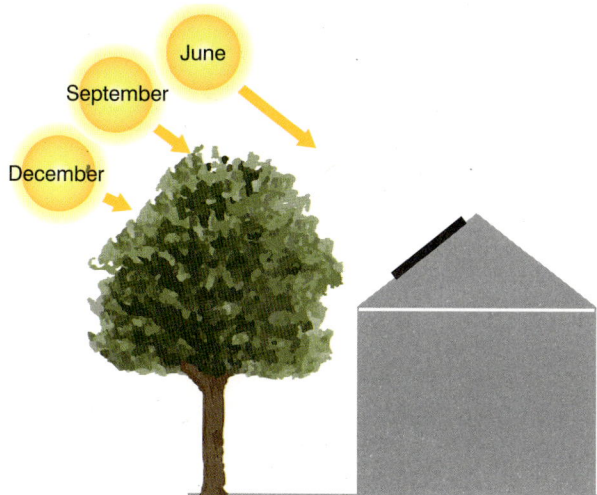

Figure 5.13 Solar PV array overshading

Heavy overshading will typically reduce the performance of the system by approximately 50 per cent during peak irradiation. Modest overshading will typically reduce performance by approximately 20 per cent.

Overshading can also lead to thermal stress in solar PV modules, causing malfunctioning to occur, possibly leading to early component failure.

The availability of a suitable solar PV array mounting structure

The solar PV array mounting structure must be suitable in terms of being:

- of sufficient size (m²) – the required area will vary according to the module efficiency. Typically, a minimum of 8 m² of suitable array mounting area is needed for each 1000 W of electricity generation under peak conditions (1 kWp or kilowatt peak).
- strong enough to support the array – factors such as imposed loads and wind uplift loads must be considered.
- in good condition – any roof or structure repairs or refurbishment should be carried out prior to installing the array.

Advantages and disadvantages of solar PV systems

Table 5.3 gives some typical advantages and disadvantages of a solar PV system.

Advantages	Disadvantages
It is a zero carbon technology	Requires a relatively large array area to make the installation worthwhile
The technology qualifies for Feed-in Tariff payments	Initial installation costs can be off-putting
Most buildings are suitable for the technology	Variable performance according to the availability of solar energy
Improves Energy Performance Certificate ratings	Some people consider that photovoltaic arrays reduce the attractiveness of buildings

Table 5.3 Typical advantages and disadvantages of a solar PV system

Micro-wind turbine systems

Introduction

The UK wind energy resource is one of the largest in Europe. This is mainly due to the fact that the UK has long exposed coastlines and low mountain ranges. In simple terms, micro-wind turbine systems are a zero carbon technology that makes use of the natural wind resource to generate electrical energy.

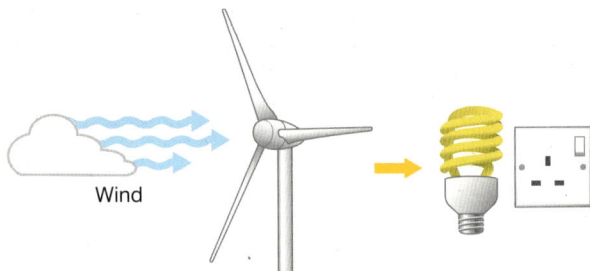

Figure 5.14 Basic principles of micro-wind turbine technology

A basic wind turbine (Figure 5.15) operates on the principle that wind passing across the rotor blades of a turbine cause a 'lift' and 'drag' effect which in turn causes the hub to rotate. The hub is connected by a low-speed shaft to a gearbox which increases the speed of rotation of the shaft. The high-speed shaft is connected to a generator that produces electricity.

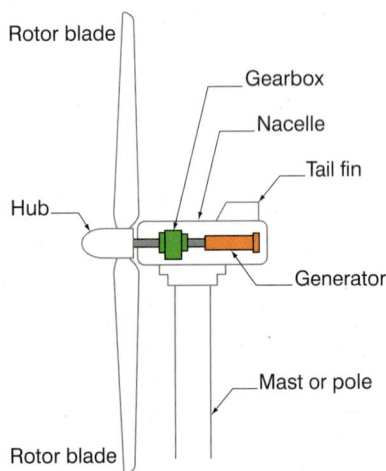

Figure 5.15 Example of a basic wind turbine (horizontal axis)

Basic system categories

Although there are a number of system types, variations and configurations, micro-wind turbine systems fall into two basic system categories:

1 on-grid systems
2 off-grid systems.

On-grid systems

On-grid systems allow any surplus electricity that is generated to be exported to the electricity distribution grid.

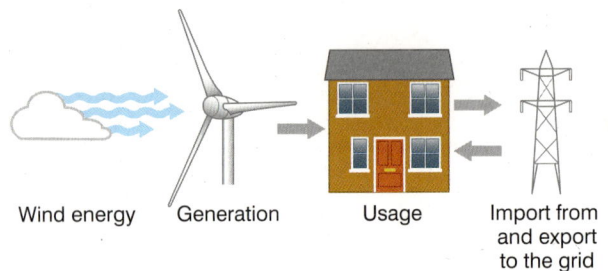

Figure 5.16 Basic principles of an on-grid micro-wind turbine system

Off-grid systems

Off-grid systems use a battery bank arrangement to store the electrical power generated for use when needed.

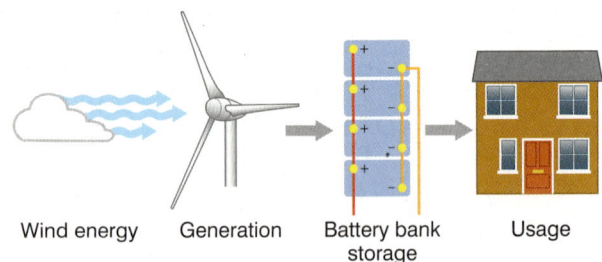

Figure 5.17 Basic principles of an off-grid micro-wind turbine system

Some micro-wind turbine systems combine on and off-grid arrangements.

The popularity of on-grid systems has increased significantly since the introduction of the Feed-in Tariff scheme.

Overview of system components

Please note that due to the intended purpose of these materials, some system components are not shown. This is not an installation diagram.

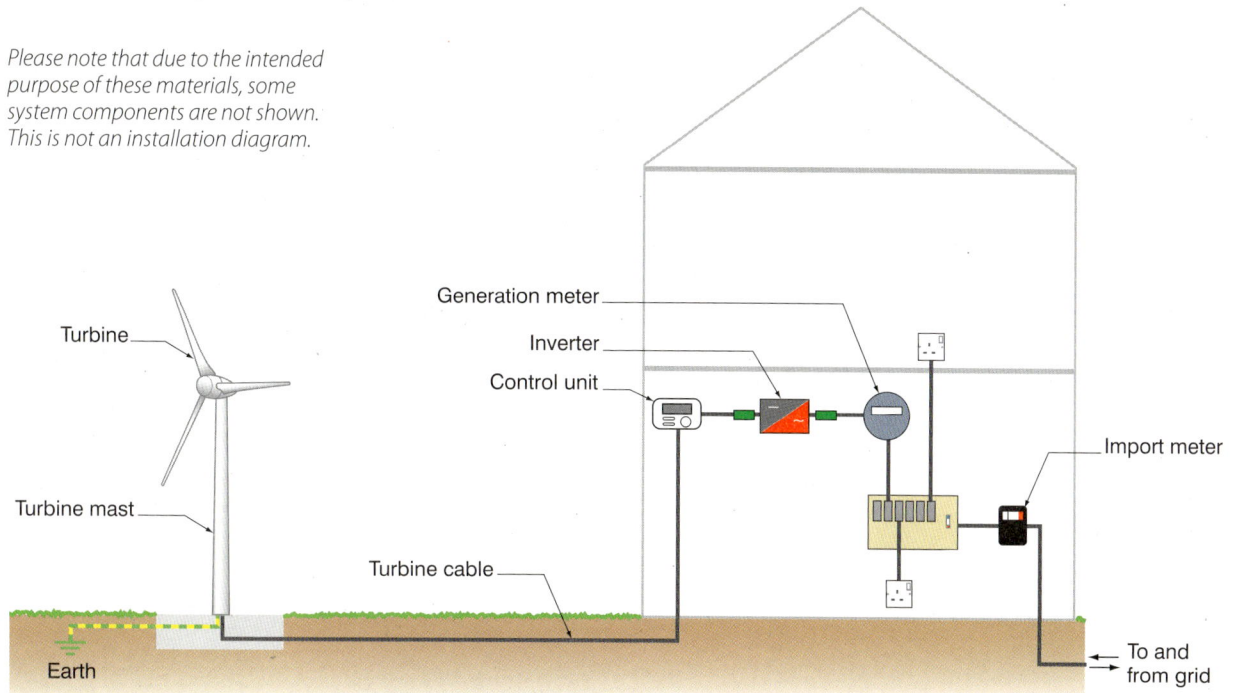

Figure 5.18 Micro-wind turbine system component overview (on-grid system)

Turbine

As outlined in the introduction, the turbine is the electricity generating component within the system. There are numerous types of wind turbine; some basic and some quite complex. There are two categories of wind turbine:

1 horizontal axis wind turbines (HAWT)
2 vertical axis wind turbines (VAWT).

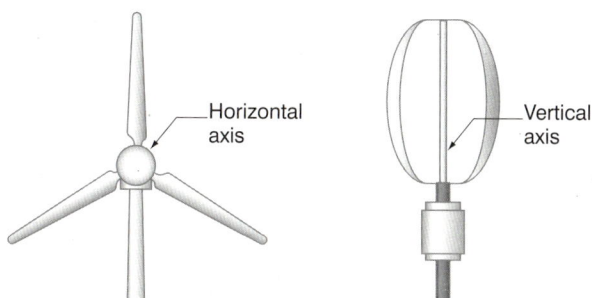

Figure 5.19 Example of a horizontal axis wind turbine (HAWT) and vertical axis wind turbine (VAWT)

HAWT include a tail fin or a yaw drive to ensure that the rotor blades are facing the direction of the prevailing wind. Vertical axis wind turbines accept wind from any direction.

Turbine mast

Wind turbines are mounted on poles or masts (sometimes referred to as towers). A number of options are available including:

- self-supporting masts
- rigid masts supported by guy ropes
- hinged masts supported by guy ropes
- building mounted poles.

Masts supported by guy ropes are typically cheaper to purchase than self-supporting masts; however, guyed masts require more space to accommodate the guy ropes. Hinged masts can be raised and lowered for easy maintenance and repair.

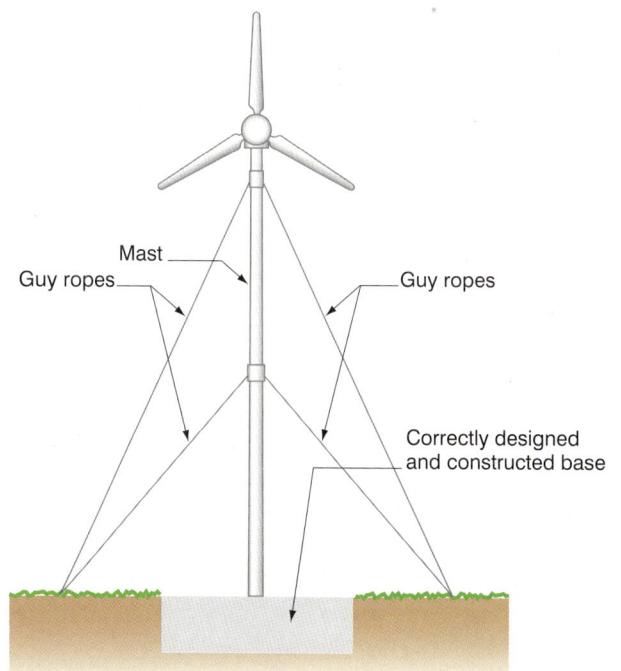

Figure 5.20 Example of a turbine mast installation supported by guy ropes

© National Skills Academy for Environmental Technologies 2011

Building mounted poles are only suitable for small micro-wind turbines and the suitability of the structure and surrounding environment for a building mounted turbine must be fully assessed.

Inverter

Most micro and small-scale wind turbines of less than 20 kW output produce 'wild' (variable voltage and frequency) alternating current (a.c.) electricity which is rectified to direct current (d.c.) via a system controller.

The d.c. electricity is then either directly used to charge batteries (off-grid system) or is converted to normal a.c. electricity (230 V, 50 Hz) using an inverter for use in an on-grid system.

Generation meter

As with Solar PV, a generation meter is fitted to record how much electricity has been generated and how much has been exported to the supply grid. In the UK, the Office of the Gas and Electricity Markets (OFGEM) has a list of approved generation meters.

Regulatory requirements relating to micro-wind turbine systems

The installation of a micro-wind turbine system will require compliance with a number of regulatory requirements including health and safety and electrical regulations. A competent installation contractor will have a detailed knowledge of these regulations and will ensure compliance.

Within this section we consider two primary regulatory requirements in relation to micro-wind turbine systems:

1 building regulations
2 planning regulations.

Building regulations

Building regulations set standards for design and construction which apply to most new buildings and many alterations to existing buildings. There are separate building regulations for:

- England and Wales
- Northern Ireland
- Scotland.

The building regulations contain various sections, a number of which have relevance or might have relevance to micro-wind turbine system installation work as summarised in Table 5.4.

Section of regulations	Relevance or potential relevance to micro-wind turbine system installation work
Structure	Where micro-wind turbines are mounted on buildings and put load on the structure
Fire safety	Where holes for cables, etc. might reduce the fire resistant integrity of the building structure
Site preparation and resistance to moisture	Where holes for cables, etc. might reduce the moisture resistant integrity of the building structure
Resistance to the passage of sound	Where holes for cables, etc. might reduce the sound proof integrity of the building structure
Electrical safety in dwellings	Safe installation of electrical controls and components

Table 5.4 Building regulations that have relevance or might have relevance to micro-wind turbine system installation work

Town and country planning regulations

As with the building regulations, there are separate town and country planning regulations for:

- England and Wales
- Northern Ireland
- Scotland.

The Town and Country Planning (General Permitted Development) Order 1995 and subsequent amendments, allow local authority planning departments to grant certain 'permitted development' rights.

The installation of micro-wind turbines is *not* classed as permitted development.

At present, planning permission is nearly always required to install a micro-wind turbine to a building, or within grounds surrounding a building.

Factors that may affect whether permission is granted or not include:

- visual impact
- noise and flicker
- vibration
- electrical interference (with TV aerials)
- safety.

Building location and feature requirements for the potential to install a micro-wind turbine system

The following building and location factors will need to be considered:

- average wind speed
- height at which the turbine can be mounted
- obstructions and turbulence

- a location that will not be affected by turbine noise, vibration and shadow flicker.

Average wind speed

The average wind speed is a critical factor in determining the viability of a micro-wind turbine system. Wind speed is measured in metres per second (m/s).

Whilst micro-wind turbines will typically start generating electricity at 3–4 m/s, the minimum viable wind speed is 5 m/s.

Most micro-wind turbines will reach their maximum rated output at between 10–14 m/s so this is the ideal wind speed range.

Wind speed can be measured on site using an anemometer but if this is done measurements should be taken over a period of months to increase accuracy.

A national wind speed database exists, but this database is no longer being updated. The database also has limitations in terms of its relevance to micro-wind turbine installations.

Height, obstructions and turbulence

As wind speed typically increases with height, the basic principle is the higher the mounting location the better. A high mounting location with a smooth prevailing wind flow is ideal. To minimise the effect of turbulence, micro-wind turbines should ideally be mounted at a distance from the nearest obstruction equal to 10 × the height of the obstruction.

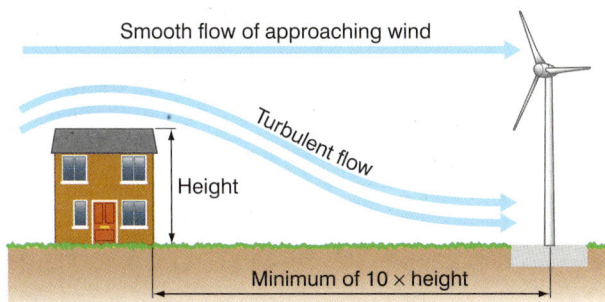

Figure 5.21 Example of wind flow behaviour and turbine mounting location

Turbine noise, vibration and shadow flicker

All wind turbines will generate some degree of noise, vibration and shadow flicker (caused by the sun passing across the turbine rotor blades as it spins). These factors are much less of a consideration when the micro-wind turbine can be located away from buildings. Where a micro-wind turbine is to be building mounted, careful consideration must be given to these factors.

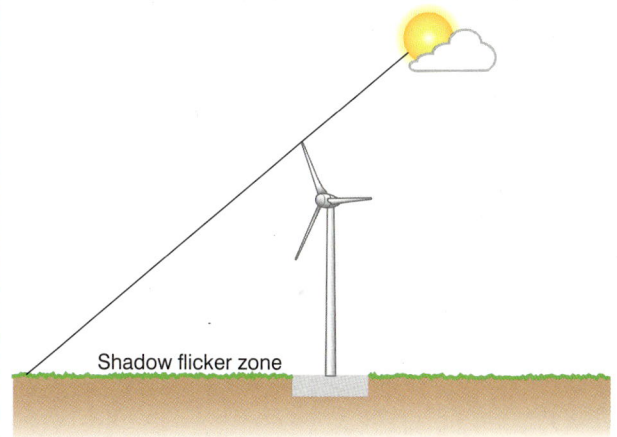

Figure 5.22 Example of a shadow flicker zone

Advantages and disadvantages of micro-wind turbine systems

Table 5.5 gives some typical advantages and disadvantages of a micro-wind turbine system.

Advantages	Disadvantages
Micro-wind turbine electricity generation output levels can be very good in the UK which has 40% of Europe's wind resource	Variable performance according to the availability of wind
The technology qualifies for Feed-in Tariff payments	Initial installation costs can be off-putting
Can be a very effective technology where no mains electricity is available	Micro-wind turbines can cause noise, vibration and shadow flicker problems
It is a zero carbon technology	Requires a suitable mounting site, ideally well away from buildings and obstructions

Table 5.5 Typical advantages and disadvantages of a micro-wind turbine system

Micro-hydropower systems

Introduction

Micro-hydropower systems make use of natural water resources to generate electrical energy.

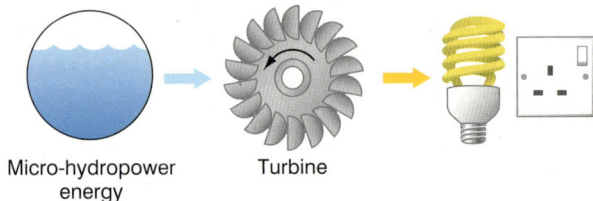

Figure 5.23 Basic principles of a micro-hydropower system

A micro-hydropower system is either installed within, or near to a suitable watercourse such as a river. Water passes through or across a turbine causing the turbine to rotate. The turbine shaft is connected to a generator and as the turbine rotates the hydropower energy is converted to electrical energy. As the water leaves the turbine it is returned to the watercourse.

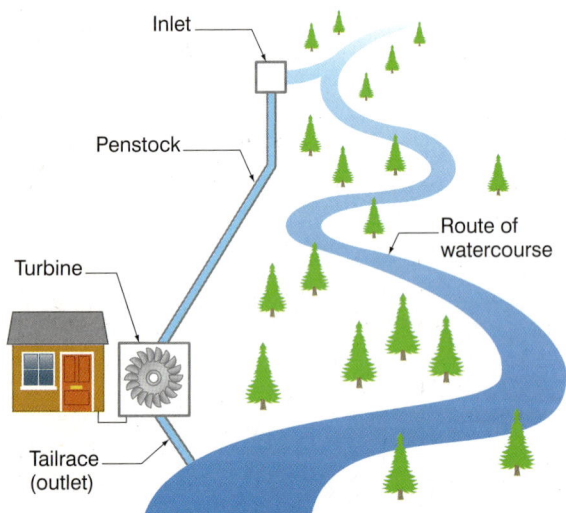

Figure 5.24 Example of a micro-hydropower system layout

When a micro-hydropower system is installed near to rather than within a suitable watercourse, a penstock is used to convey the water from the inlet to the turbine. A penstock is a pipe. A tailrace is used to return the water to the watercourse. A tailrace can be a pipe or a channel.

Micro-hydropower is a zero carbon technology.

Basic system categories

As with other electricity producing technologies there are a number of system types, variations and configurations.

Micro-hydropower turbine systems fall into two basic system categories:

1 on-grid systems
2 off-grid systems.

On-grid systems

As with Solar PV and micro-wind systems, on-grid micro-hydropower systems allow any surplus electricity that is generated to be exported to the electricity distribution grid. This type of system is included in the Feed-in Tariff scheme.

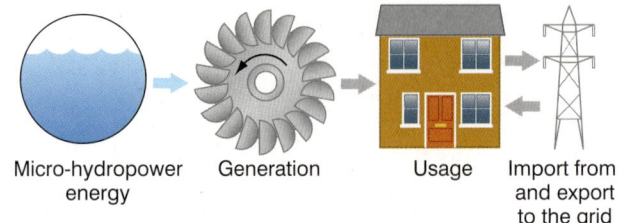

Figure 5.25 Basic principles of an on-grid micro-hydropower system

Off grid systems

Off-grid systems use a battery bank arrangement to store the electrical power generated for use when needed.

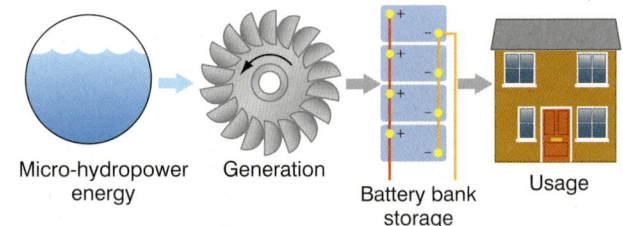

Figure 5.26 Basic principles of an off-grid micro-hydropower system

Some micro-hydropower systems combine on and off-grid arrangements.

As with other electricity producing technologies, micro-hydropower systems will include an inverter to convert d.c. electricity to a.c. electricity and on-grid systems will also include a generation meter.

Building location and feature requirements for the potential to install a micro-hydropower system

The following building and location factors will need to be considered:

- the availability of a water course (river, stream, etc.)
- the ability to achieve adequate 'hydraulic head' and 'flow' within the system design
- a suitable location for an inlet
- a suitable location for the turbine and generator
- a suitable location for the tailrace outlet.

Hydraulic head and flow

Head is the vertical distance (measured in metres) between the upper and lower water levels or the vertical distance between the intake and turbine.

Flow is the quantity of water, measured in metres cubed per second (m^3/s), that is moving over a given period of time.

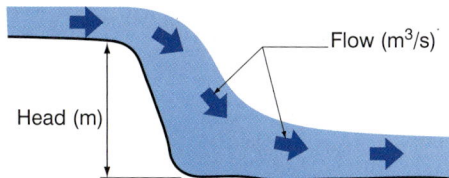

Figure 5.27 Head and flow in a watercourse with a weir

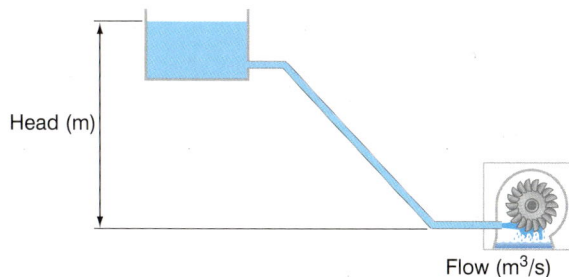

Figure 5.28 Head and flow in an installation with a penstock

The available hydraulic head and flow is also a key factor that will determine the type of micro-hydropower turbine that can be used.

Micro-hydropower turbines

Turbine classification

Micro-hydropower turbines are classified according to their ability to operate in high, medium or low head conditions. They are also classified as being either impulse turbines or reaction turbines according to how they operate.

- **Impulse turbine** – the turbine wheel or runner is surrounded by air and the turbine is moved by the impulse created by a jet or jets of water that are aimed at the turbine. Types of impulse turbine include Pelton, Multi-jet Pelton, Turgo and Crossflow.
- **Reaction turbine** – the turbine wheel or runner is fully immersed in water and the turbine is moved in reaction to the flow of water. Types of reaction turbine include Francis (spiral case), Francis (open-flume), Propeller and Kaplan.

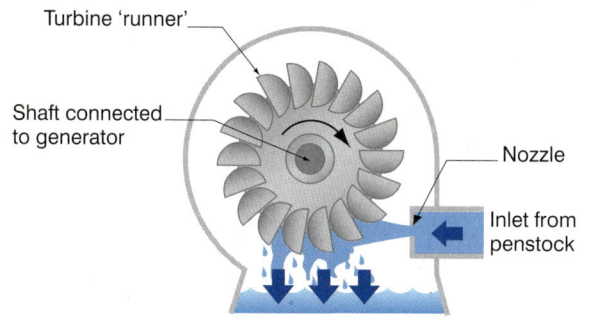

Figure 5.29 Example of an impulse turbine (Pelton type)

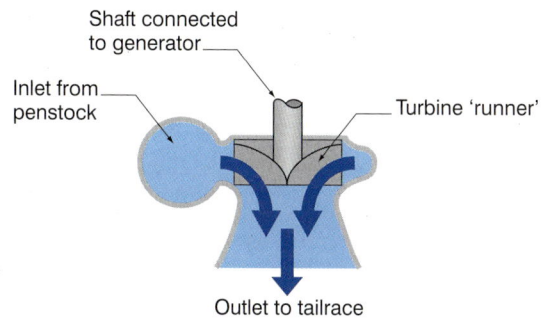

a) Section through turbine from the side

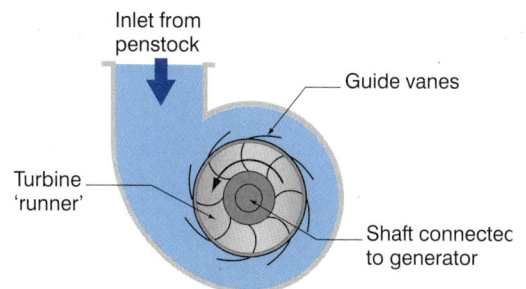

b) Section through turbine from above

Figure 5.30 Example of a reaction turbine (horizontal Francis type)

Turbine applications in relation to the available head of water

Turbine type	Head classification		
	High (>50 m)	Medium (10–50 m)	Low (<10m)
Impulse	Pelton Turgo Multi-jet Pelton	Crossflow Turgo Multi-jet Pelton	Crossflow
Reaction		Francis (spiral case)	Francis (open-flume) Propeller Kaplan

Table 5.6 Turbine applications in relation to the available head of water

In addition to the impulse and reaction turbines described above, an alternative type of turbine that is sometimes used for larger micro-hydropower schemes is the reverse Archimedean screw turbine. This type of turbine is particularly suitable for low head installations and is also fish friendly allowing fish and eels to pass through without injury.

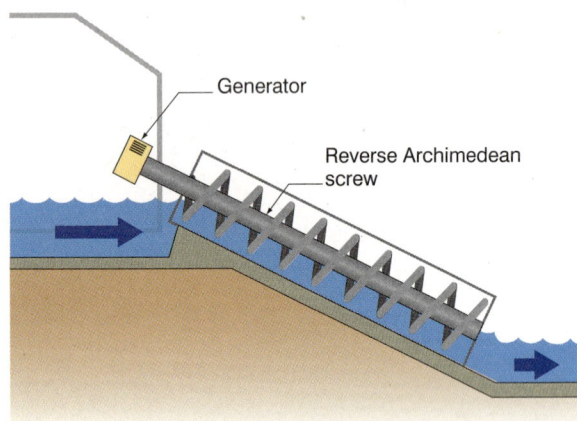

Figure 5.31 Example of a reverse Archimedean screw turbine

Figure 5.32 Reverse Archimedean screw turbines being installed

Regulatory requirements relating to micro-hydropower systems

The installation of a micro-hydropower system will require compliance with a number of regulatory requirements including health and safety, water and electrical regulations. A competent installation contractor will have a detailed knowledge of these regulations and will ensure compliance.

Within this section we consider three primary regulatory requirements in relation to micro-hydropower systems:

1 building regulations
2 planning regulations
3 environmental regulations.

Building regulations

Building regulations set standards for design and construction which apply to most new buildings and many alterations to existing buildings. There are separate building regulations for:

- England and Wales
- Northern Ireland
- Scotland.

The building regulations contain various sections, a number of which have relevance or might have relevance to micro-hydropower system installation work as summarised in Table 5.7.

Section of the regulations	Relevance or potential relevance to micro-hydropower system installation work
Structure	Where any part of the micro-hydropower system puts load on the structure
Fire safety	Where holes for cables, etc. might reduce the fire resistant integrity of the building structure
Site preparation and resistance to moisture	Where holes for cables, etc. might reduce the moisture resistant integrity of the building structure
Resistance to the passage of sound	Where holes for cables, etc. might reduce sound proof integrity of the building structure
Electrical safety in dwellings	Safe installation of electrical controls and components in dwellings

Table 5.7 Building regulations that have relevance or might have relevance to micro-hydropower system installation work

Town and country planning regulations

As with the building regulations, there are separate town and country planning regulations for:

- England and Wales
- Northern Ireland
- Scotland.

The key features of a micro-hydropower scheme include:

- a hydraulic head (vertical distance from water source to the turbine)
- a water intake
- a pipe or channel to take water to the turbine
- a turbine, generator and electrical connection
- an outflow, where the water returns to the watercourse.

These elements raise a number of important planning issues and planning permission will usually be needed. The elements of a micro-hydropower scheme create potential impacts on:

- landscape and visual amenity
- nature conservation
- the water regime.

Environmental regulation

All water courses in England and Wales are controlled by the Environment Agency and in Scotland by the Scottish Environment Protection Agency. Advice should be sought from these agencies prior to any installation.

To remove water (even though it might be returned) from a water course of any size in England and Wales will almost certainly require permission from the Environment Agency in the form of a licence.

There are three licences that can apply to a micro-hydropower scheme:

1 **abstraction licence** – if water is being diverted away from the main line of flow of the river part of the consideration will be fish migration. Most micro-hydropower turbines are not fish friendly and where fish migration is a factor, an abstraction licence will only be issued with conditions stating the requirement for fish screens and a fish pass arrangement

2 **impoundment licence** – if changes are being made to structures which impound water (such as weirs and sluices) or if new structures are to be built
3 **land drainage consent** – for any works being carried out in a main channel.

It is necessary to carry out an environmental site audit (ESA) as part of the process of identifying the suitability of a micro-hydropower installation. The ESA covers the following areas:

- water resources
- conservation
- chemical and physical water quality
- biological water quality
- fisheries
- flood risk
- navigation.

The Environment Agency must always be consulted as early as possible when a micro-hydropower installation is being considered.

Advantages and disadvantages of micro-hydropower systems

Table 5.8 gives some typical advantages and disadvantages of a micro-hydropower system.

Advantages	Disadvantages
It is a zero carbon technology	Requires a watercourse with suitable head and flow
The technology qualifies for Feed-in Tariff payments	Initial installation costs can be off-putting
Excellent payback potential	Usually requires planning permission
Can be a very effective technology where no mains electricity is available	Requires permission from the Environment Agency

Table 5.8 Typical advantages and disadvantages of a micro-hydropower system

Section six: Co-generation technologies

Micro-combined heat and power systems (heat-led)

Introduction

A heat-led micro-combined heat and power (micro-CHP) system includes a micro-CHP unit, similar in appearance to a heating system boiler, that generates some electricity as well as generating heat for domestic hot water and space heating purposes.

Although micro-CHP units have existed for some time, units suitable for domestic installations have only recently become available. The currently available domestic units are gas-fired only. Other fuels options may be available for non-domestic units.

The term 'heat led' means that the generation of electricity occurs when the unit is responding to a system demand for heat and that the majority of output from the unit is for heating purposes.

Micro-CHP units are typically up to 95 per cent efficient with energy flows of approximately 80 per cent heat output and 15 per cent electrical output.

Micro-CHP is a low carbon technology. micro-CHP installations are eligible for Feed-in Tariff payments providing the installation is carried out by a Microgeneration Certification Scheme (MCS) certified contractor using an MCS approved unit.

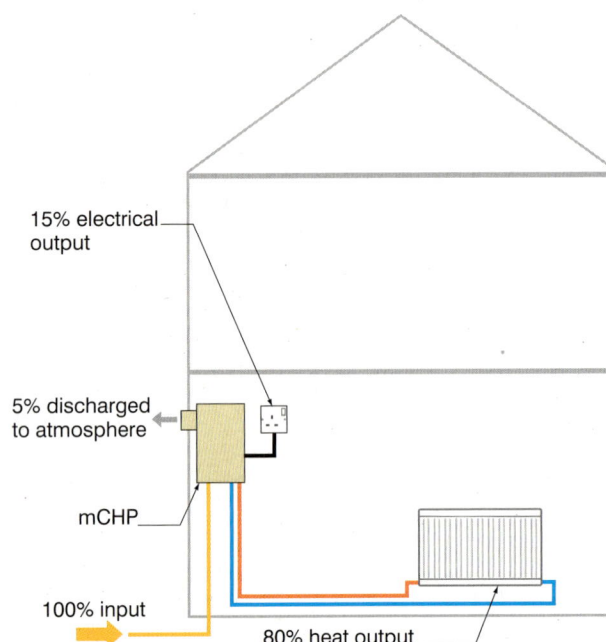

15% electrical output

5% discharged to atmosphere

mCHP

100% input

80% heat output

Figure 6.1 Typical micro-CHP system energy flows

Micro-combined heat and power units

The currently available heat-led micro-CHP units are engine or turbine based units. Fuel cell micro-CHP units are in development and some manufacturers are currently undertaking field trials for their fuel cell micro-CHP units. Currently available micro-CHP units will contain one of the following engine types:

- external combustion
- internal combustion
- organic rankine cycle.

The internal components of a key engine or turbine based micro-CHP unit are:

- an engine or gas turbine
- an alternator
- two heat exchangers
- a supplementary burner
- a combustion fan
- electrical controls.

Domestic micro-CHP units typically include a Stirling type external combustion engine.

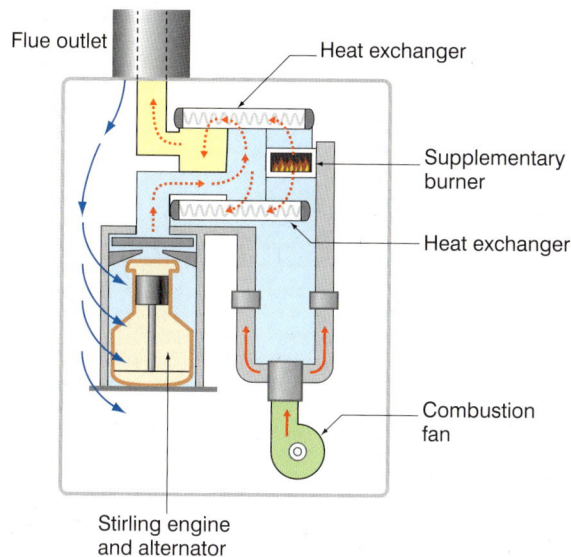

Figure 6.2 Example of a micro-CHP unit with a Stirling engine

A Stirling engine micro-CHP unit operates in the following sequence:

1 When demand for heat occurs, a gas burner provides heat to the Stirling engine unit causing the Stirling engine to operate.
2 The Stirling engine unit includes a generator comprising a piston that moves between a copper coil. As the Stirling engine operates, electricity is generated providing the engine runs for a minimum period of time and does not cycle on and off.
3 There is a limit (typically 25 per cent of total unit output) to the amount of heat that can be provided during the operation of the Stirling engine.
4 When additional heat is needed to meet higher demand, the supplementary burner operates.

Electrical output and system connections

A domestic micro-CHP unit will typically generate between 1 kW and 1.5 kW of electricity. Larger micro-CHP units typically generate up to 5–6 kW of electricity. The preferred connection arrangement between the micro-CHP unit and the main electrical installation is to use a dedicated circuit from/to the consumer unit.

Where this is difficult, it is possible to connect the unit to an existing final circuit.

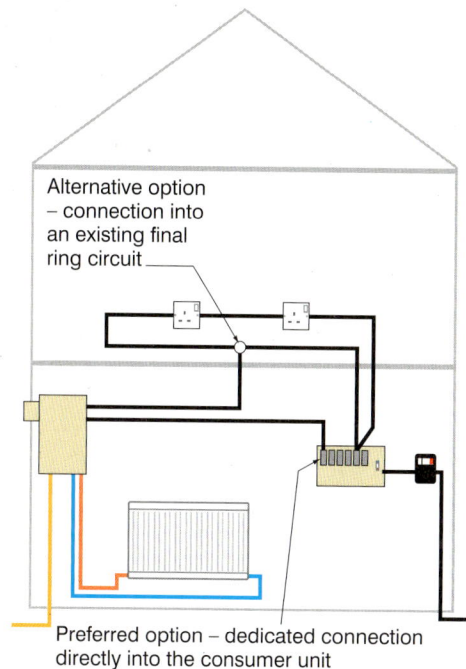

Figure 6.3 Micro-CHP unit electrical connection options

All electrical work must be designed, installed and tested by a competent person.

Any surplus electricity can be exported to the distribution grid.

Regulatory requirements relating to micro-CHP systems (heat-led)

The installation of a micro-CHP system will require compliance with a number of regulatory requirements including health and safety, water and electrical regulations. A competent installation contractor will have a detailed knowledge of these regulations and will ensure compliance.

Within this section we consider two primary regulatory requirements in relation to micro-CHP systems:

1 building regulations
2 planning regulations.

Building regulations

Building regulations set standards for design and construction which apply to most new buildings and many alterations to existing buildings. There are separate building regulations for:

- England and Wales
- Northern Ireland
- Scotland.

The building regulations contain various sections, a number of which have relevance or might have relevance to micro-CHP system installation work as summarised in Table 6.1.

Section of Regulations	Relevance or potential relevance to micro-CHP system installation work
Structure	Where the micro-CHP unit and other components put load on the structure
Fire safety	Where holes for pipes, etc. might reduce the fire resistant integrity of the building structure
Site preparation and resistance to moisture	Where holes for pipes, etc. might reduce the moisture resistant integrity of the building structure
Resistance to the passage of sound	Where holes for pipes, etc. might reduce sound proof integrity of the building structure
Sanitation, hot water safety and water efficiency	Hot water safety and water efficiency
Combustion appliances and fuel storage system	Micro-CHP units are a heat-producing combustion appliances and must be installed safely
Conservation of fuel and power	Energy efficiency of the system and the building
Electrical safety in dwellings	Safe installation of electrical controls and components

Table 6.1 Building regulations that have relevance or might have relevance to micro-CHP system installation work

Town and country planning regulations

As with the building regulations, there are separate town and country planning regulations for:

- England and Wales
- Northern Ireland
- Scotland.

The Town and Country Planning (General Permitted Development) Order 1995 and subsequent amendments, allow local authority planning departments to grant certain 'permitted development' rights.

Planning permission is not normally needed when installing a micro-combined heat and power system in a house if the work is all internal. However, if the installation requires an outside flue it will normally be permitted development if the flue is on the rear or side elevation of the building and projects to no more than one metre above the highest part of the roof.

If the building is listed or in a designated area, even if the building has permitted development rights it is advisable to check with the local planning authority before a flue is fitted. Consent is also likely to be needed for internal alterations.

In a conservation area or World Heritage site the flue should not be fitted on the principal or side elevation if it would be visible from a highway. If the project also requires an outside building to house any of the micro-CHP equipment, the same rules apply to that building as for other extensions and garden outbuildings.

Building location and feature requirements for the potential to install a micro-CHP system (heat-led)

The following building and location factors will need to be considered:

- A suitable route and termination point for the micro-CHP unit flue system.
- A suitable heat-demand – heat-led micro-CHP units only generate electricity when the unit engine is able to run for a minimum period of time. Additionally, the unit will not be as efficient if the unit cycles on and off. Small dwellings and dwellings with low heat demand are not usually suitable for heat-led micro-CHP.

Advantages and disadvantages of micro-combined heat and power systems (heat-led)

Table 6.2 gives some typical advantages and disadvantages of a micro-CHP system.

Advantages	Disadvantages
Domestic micro-CHP units are now similar in size to a central heating boiler	The cost of domestic micro-CHP units do not compare favourably to central heating boilers
Heat-led micro-CHP units produce free electricity whilst generating heat	Heat-led micro-CHP units are not suitable for properties with low heat demand
Eligible for Feed-in Tariff payments (subject to conditions)	Heat-led micro-CHP units have a limited electrical generation capacity
Does not rely on building orientation or weather conditions to generate renewable electricity	Unlike other renewable electricity producing technologies, micro-CHP is a low carbon rather than zero carbon technology

Table 6.2 Typical advantages and disadvantages of a micro-CHP system

Section seven: Water conservation technologies

Climate change is resulting in wetter winters, rising sea levels and hotter drier summers. Wetter winters and rising sea levels have led to more frequent and widespread flooding in the UK. Hotter drier summers deplete water supplies leading to droughts and water shortages. These factors have led to the need to reduce the consumption of wholesome mains water.

The water conservation technologies included in this section are:

- rainwater harvesting
- greywater reuse.

To become 'wholesome', water is treated by the water supply undertaker before it is supplied to the consumer. The treatment process consumes energy and results in carbon emissions.

Although not typically associated with being a low carbon technology, because rainwater harvesting systems and greywater reuse reduce wholesome water consumption, the use of recycled water can also lead to energy savings and a carbon emission reduction through a reduction in wholesome treated water consumption. However, if energy is used to treat the harvested rainwater or greywater, this may offset any energy and carbon emission savings achieved through the reduction in wholesome treated water consumption.

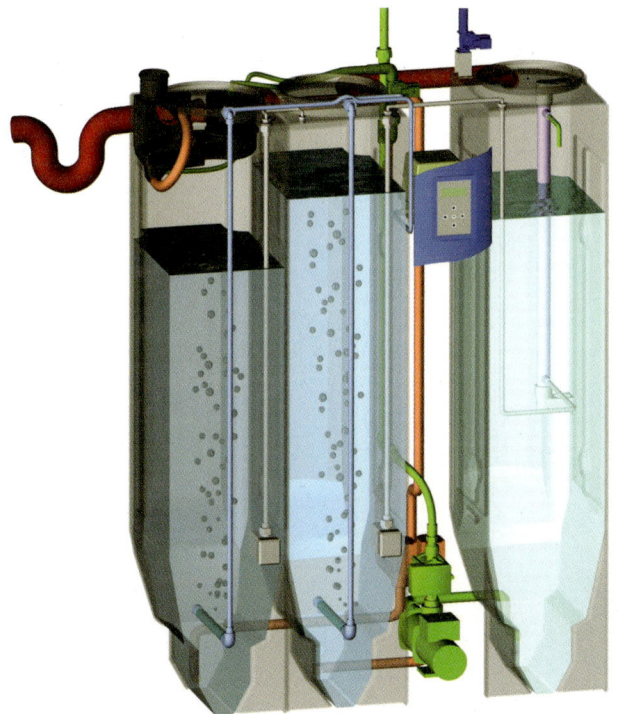

Rainwater harvesting systems

Introduction

A rainwater harvesting system captures and stores rainwater for permitted non-wholesome usage.

Harvested rainwater is classified as a class 5 risk under the Water Supply (Water Fittings) Regulations 1999. This classification is the highest risk classification given and as a result harvested rainwater can only be used for the uses given in Figure 7.1.

Figure 7.1 Permitted uses for harvested rainwater

Harvested rainwater is *not suitable* for use and is *not permitted* for use for the following purposes:

- drinking water
- dishwashing (hand or machine)
- food preparation
- personal washing, showering and bathing.

Types of rainwater harvesting system

There are three main types of rainwater harvesting system:

1 gravity fill and distribution
2 pumped to storage with gravity distribution
3 direct pumped.

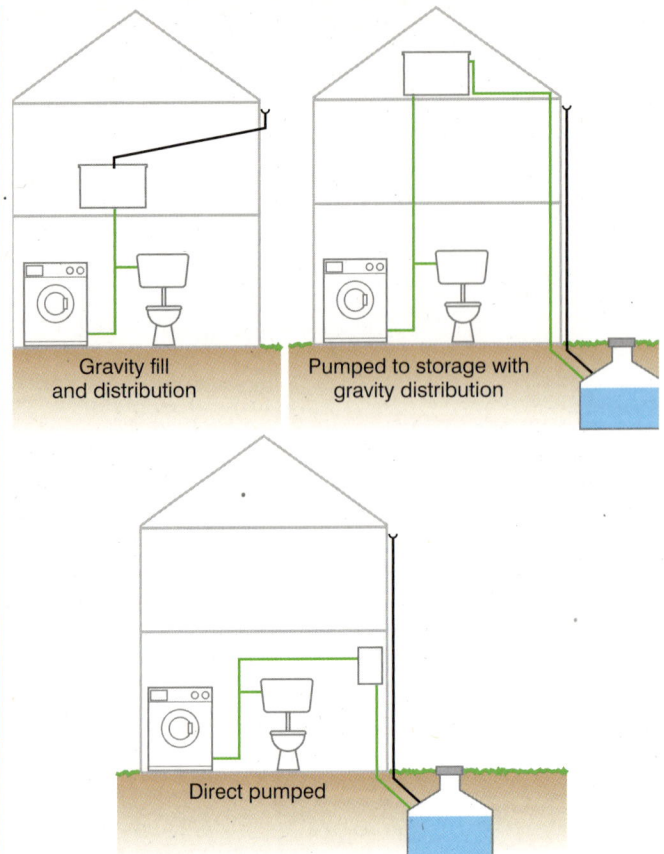

Figure 7.2 Types of rainwater harvesting system

Gravity fill and distribution

In a gravity fill and distribution system the rainwater storage tank is positioned above the appliances to be supplied with harvested rainwater and below the source of the rainwater. No pump is required as rainwater drains into the tank by gravity and the water is distributed to the appliances by gravity. This system type is not as common as pumped system types.

Pumped to storage with gravity distribution

In a pumped to storage with gravity distribution system rainwater drains into the tank by gravity and a pump, typically submersible and located within the storage tank, is used to pump the harvested rainwater to a storage cistern within the building. The stored water is then distributed to the appliances by gravity.

Direct pumped

In a direct pumped system rainwater drains into the tank by gravity and a pump, typically submersible and located within the storage tank, is used to pump the harvested rainwater directly to the appliances.

Rainwater harvesting storage tanks can be positioned externally above or below ground or inside the building.

Rainwater harvesting system components

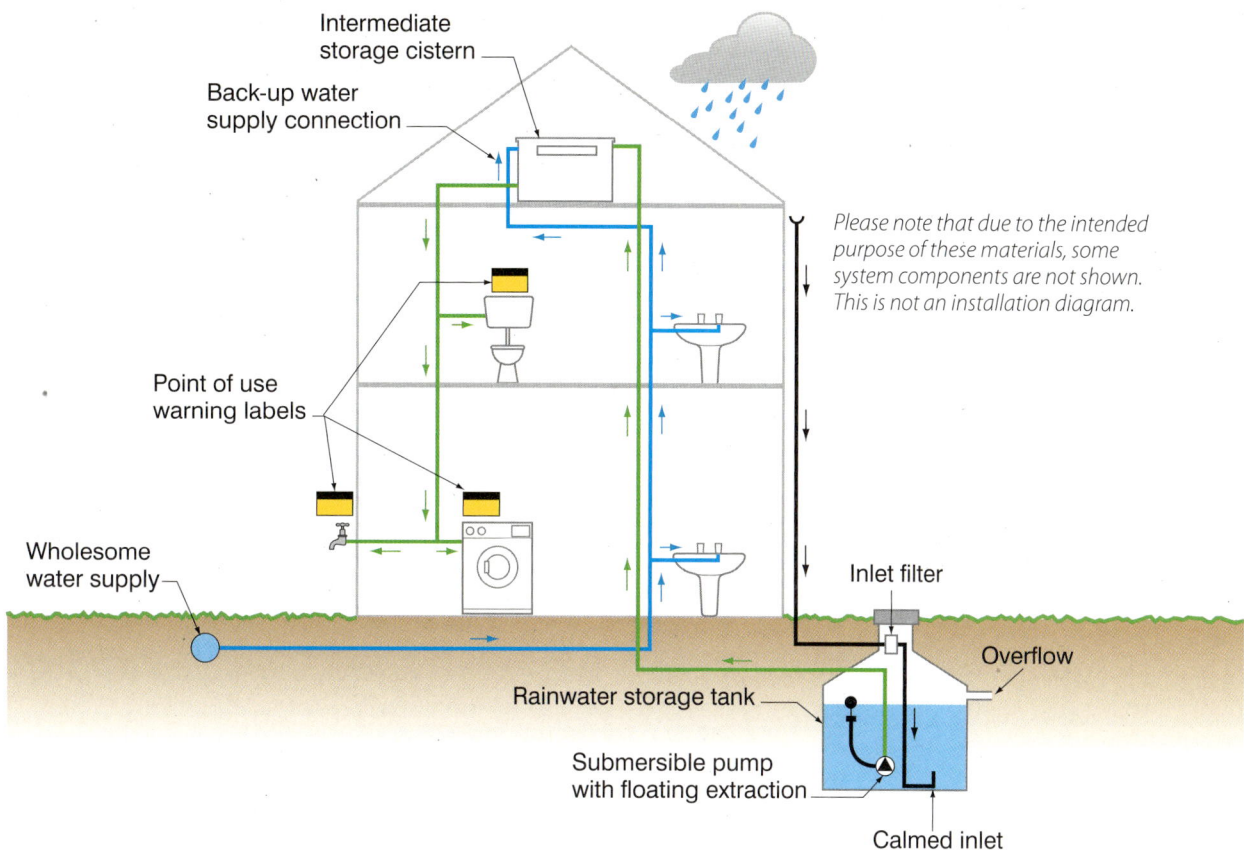

Figure 7.3 Rainwater harvesting system components *(based upon a pumped to storage with gravity distribution system)*

Key system components

Inlet filter

Before the harvested rainwater enters the storage tank it must pass through an approved type inlet filter. The inlet filter is often fitted within the tank, just below the access cover, but can be located anywhere in the collection pipework providing it is accessible for maintenance purposes.

Calmed inlet

The calmed inlet is fitted on the end of the pipe feeding into a storage tank. The purpose of the calmed inlet is to minimise turbulence and slow the water flow into the tank to avoid unnecessary disturbance of any sediment at the base of the tank.

Submersible pump with floating extraction

A submersible pump with a floating extraction point is typically located within the storage tank. The floating extraction point ensures that the pump is supplied with water near to the surface, reducing the risk of any sediment entering the pump and the distribution pipework. Submersible pumps are fitted with dry-run protection to prevent damage to the pump.

Point of use warning labels

It is a requirement of the water regulations that all points of use be labelled to identify that they are supplied with harvested rainwater. In addition, the water regulations also require that all distribution pipework is labelled to identify that it is part of a harvested rainwater system.

Back-up water supply

There is potential that in times of drought and/or heavy demand, the harvested rainwater system will not be able to supply the outlets connected to the system. For this reason, it is normal practice for a back-up water supply connection to be provided from the wholesome water supply system. As harvested rainwater is a class 5 risk, the back-up water must be connected by a suitable backflow prevention arrangement to prevent backflow of the stored rainwater into the wholesome water supply system. The only backflow prevention arrangement that is suitable for a class 5 risk is an arrangement that provides a physical air gap between the stored rainwater and the wholesome water supply connection. Type AA and Type AB air gaps are suitable backflow prevention arrangements.

47

Regulatory requirements relating to rainwater harvesting technologies

The installation of a rainwater harvesting system will require compliance with a number of regulatory requirements including health and safety, and water regulations. A competent installation contractor will have a detailed knowledge of these regulations and will ensure compliance.

Within this section we consider two primary regulatory requirements in relation to rain water harvesting systems:

1 building regulations
2 planning regulations.

Building regulations

Building regulations set standards for design and construction which apply to most new buildings and many alterations to existing buildings. There are separate building regulations for:

- England and Wales
- Northern Ireland
- Scotland.

The building regulations contain various sections, a number of which have relevance or might have relevance to rainwater harvesting system installation work as summarised in Table 7.1:

Section of regulations	Relevance or potential relevance to rainwater harvesting system installation work
Structure	Where rainwater harvesting system components put load on the structure and/or where excavations are made near to the structure
Fire safety	Where holes for pipes, etc. might reduce the fire resistant integrity of the building structure
Site preparation and resistance to moisture	Where holes for pipes, etc. might reduce the moisture resistant integrity of the building structure
Resistance to the passage of sound	Where holes for pipes, etc. might reduce the sound proof integrity of the building structure
Sanitation, hot water safety and water efficiency	Water efficiency
Drainage and waste disposal	Rainwater gutters and rainwater pipework connected to rainwater harvesting systems
Electrical safety in dwellings	The connection of rainwater harvesting system electrical components

Table 7.1 Building regulations that have relevance or might have relevance to rainwater harvesting system installation work

Town and country planning regulations

As with the building regulations, there are separate town and country planning regulations for:

- England and Wales
- Northern Ireland
- Scotland.

Planning permission is not normally needed when installing a rainwater harvesting system in a house if the finished installation does not alter the outside appearance of the property. Where above ground rainwater harvesting storage tanks are to be included, planning permission may be required.

If the building is listed or in a designated area it is advisable to check with the local planning authority before installing a rainwater harvesting system even if the building has permitted development rights. Consent is also likely to be needed for internal alterations to listed buildings.

The local planning authority should also be consulted if the property is in a conservation area or in a World Heritage site.

If the project requires an outside building to house the rainwater harvesting storage tanks, the same rules apply to that building as for other extensions and garden outbuildings.

Building location and feature requirements for the potential to install a rainwater harvesting system

For the potential to install a rainwater harvesting system, as a minimum, some or all of the following building and location factors will need to be considered:

- a suitable location and space for a storage tank of a suitable size to meet the demand
- a suitable location for rainwater harvesting system storage tank(s) to minimize the potential for freezing, warming and algal blooms
- for retrofit installations, access for excavation machinery might also need to be considered
- a suitable supply (yield) of rainwater in relation to the demand on the system. Rainwater harvesting systems are not suitable for areas with a low rainfall intensity or for buildings with a small rainwater catchment area
- the availability of a wholesome back-up water supply.

Advantages and disadvantages of rainwater harvesting systems

Table 7.2 gives some typical advantages and disadvantages of a rainwater harvesting system.

Advantages	Disadvantages
Conserves wholesome water	Payback periods can be long
Indirectly reduces energy consumption and reduces carbon emissions	Not always straightforward to install to an existing building
A wide range of system options exist	There is a risk of contamination or cross-connection
Rainwater is free, so for buildings where a water meter is fitted the annual cost of water will reduce	Only certain types of outlet and appliance can be supplied using harvested rainwater

Table 7.2 Typical advantages and disadvantages of a rainwater harvesting system

Greywater reuse systems

A greywater reuse system receives 'grey' waste water that is discharged from washbasins, baths, showers and washing machines so that the greywater can be reused at permitted non-wholesome outlets.

Like harvested rainwater, greywater is classified as a class 5 risk under the Water Supply (Water Fittings) Regulations 1999 and as a result greywater can only be reused for the purposes given in Figure 7.4.

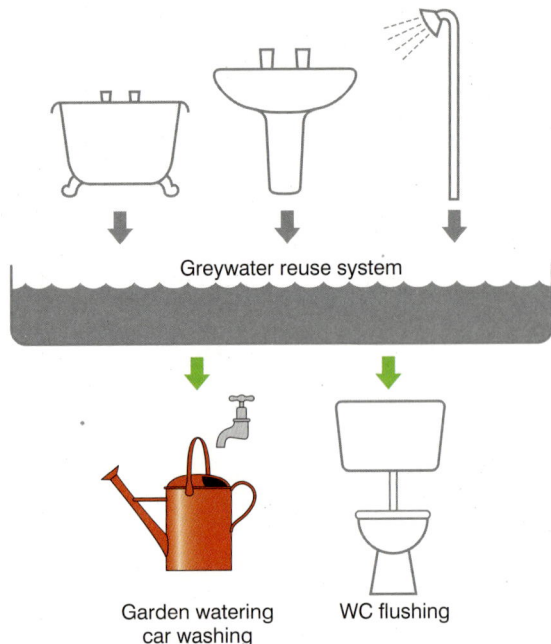

Figure 7.4 Permitted uses for greywater

In some circumstances, it is possible to use greywater to supply a clothes washing machine if the greywater is appropriately treated. If greywater is to be used for irrigation, it should be directly applied to soil and not through a sprinkler or method that would allow contact with above ground portions of plants. Greywater should not be used to water crops, which are eaten uncooked. It is recommended that greywater should not be applied to seedlings or young plants.

Greywater is *not suitable* for use and is *not permitted* for use for the following purposes:

- drinking water
- dishwashing (hand or machine)
- food preparation
- personal washing, showering and bathing.

Greywater reuse system types

There are six main types of greywater reuse system. Each type is identified and briefly described in Table 7.3.

System type	Description
Direct reuse system	A system that collects greywater from appliances and delivers it directly to the points of use with no treatment and minimal or no storage
Short retention system	A system that includes a basic filtration or treatment technique such as surface skimming and allows for natural particle settlement
Basic physical/ chemical system	A system that filters greywater prior to storage and uses chemical disinfectants such as chlorine or bromine to stop bacterial growth during storage
Biological system	A system that introduces an agent, such as oxygen, into the stored greywater to allow bacteria to digest any organic matter. Pumps or plants can be used to aerate the stored water
Biomechanical system	A system that combines both physical and biological treatment
Hybrid system	A combination of any of the above systems or a combined rainwater harvesting and greywater reuse system

Table 7.3 Types of greywater reuse system

The type of system selected will be influenced by a number of factors including the quality of the greywater being received and the intended reuse purposes.

The most common greywater reuse system is direct pumping from the storage tank to the greywater outlets. An example of a direct pumped greywater reuse system is shown in Figure 7.5.

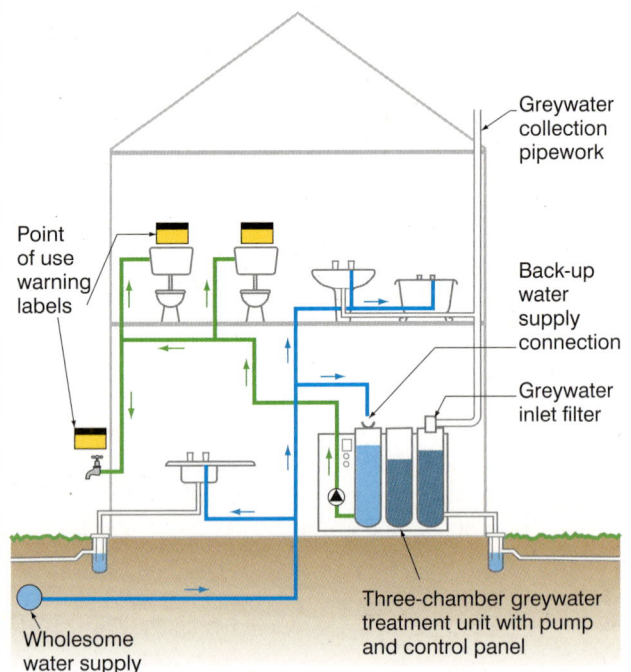

Figure 7.5 Example of a greywater reuse system (direct pumped)

The greywater treatment unit illustrated in Figure 7.5 uses physical, biological and UV disinfection to treat and prepare the greywater for reuse.

Water enters the unit via an inlet filter, biological treatment takes place in chambers 1 and 2, and UV disinfection takes place in chamber 3.

As with rainwater harvesting systems, all greywater reuse system supply points and pipework must be marked and labelled to minimise the risk of incorrect use of reclaimed greywater and/or the possibility of cross connection between wholesome water systems and greywater reuse systems.

Regulatory requirements relating to greywater reuse technologies

The installation of a greywater reuse system will require compliance with a number of regulatory requirements including health and safety, and water regulations. A competent installation contractor will have a detailed knowledge of these regulations and will ensure compliance.

Within this section we consider two primary regulatory requirements in relation to greywater reuse systems:

1 building regulations
2 planning regulations.

Building regulations

Building regulations set standards for design and construction which apply to most new buildings and many alterations to existing buildings. There are separate building regulations for:

- England and Wales
- Northern Ireland
- Scotland.

The building regulations contain various sections, a number of which have relevance or might have relevance to greywater reuse system installation work as summarised in Table 7.4.

Section of regulations	Relevance or potential relevance to greywater reuse system installation work
Structure	Where greywater reuse system components put load on the structure and/or where excavations are made near to the structure
Fire safety	Where holes for pipes, etc. might reduce the fire resistant integrity of the building structure
Site preparation and resistance to moisture	Where holes for pipes, etc. might reduce the moisture resistant integrity of the building structure
Resistance to the passage of sound	Where holes for pipes, etc. might reduce the sound proof integrity of the building structure
Sanitation, hot water safety and water efficiency	Water efficiency
Drainage and waste disposal	Sanitary pipework connected to greywater reuse systems
Electrical safety in dwellings	The connection of greywater reuse system electrical components

Table 7.4 Building regulations that have relevance or might have relevance to greywater reuse system installation work

Town and country planning regulations

As with the building regulations, there are separate town and country planning regulations for:

- England and Wales
- Northern Ireland
- Scotland.

Planning permission is not normally needed when installing a greywater reuse system in a house if the finished installation does not alter the outside appearance of the property. Where above ground greywater reuse storage tanks are to be included, planning permission might be required.

If the building is listed or in a designated area it is advisable to check with the local planning authority before installing a greywater reuse system even if the building has permitted development rights. Consent is also likely to be needed for internal alterations to listed buildings.

The local planning authority should also be consulted if the property is in a conservation area or in a World Heritage site.

If the project requires an outside building to house the greywater reuse storage tanks the same rules apply to that building as for other extensions and garden outbuildings.

Building location and feature requirements for the potential to install a greywater reuse system

The following building and location factors will need to be considered:

- a suitable location and space for a storage tank of a suitable size to meet the demand
- a suitable location for greywater system storage tank(s) to minimize the potential for freezing, warming and algal blooms
- for retrofit installations, access for excavation machinery might also need to be considered
- a suitable supply (yield) of greywater in relation to the demand on the system. Greywater reuse systems are not suitable for buildings with a low volume of greywater discharge.

Advantages and disadvantages of greywater reuse systems

Table 7.5 shows some typical advantages and disadvantages of a greywater reuse system.

Advantages	Disadvantages
Conserves wholesome water	Payback periods can be long
Indirectly reduces energy consumption and reduces carbon emissions	Not always straightforward to install to an existing building
A wide range of system options exist	There is a risk of contamination or cross-connection
Greywater is free, so for buildings where a water meter is fitted the annual cost of water will reduce	Only certain types of outlet and appliance can be supplied using greywater

Table 7.5 Typical advantages and disadvantages of a greywater reuse system

Glossary

Term	System context	Meaning
Algal bloom	Rainwater harvesting Greywater reuse	A rapid increase or accumulation in the population of algae.
Auger feed mechanism	Biomass	A screw type automated fuel feed arrangement designed to feed the required volume of fuel at a consistent rate.
Auxillary heat source system	Heat pumps	A heat source system intended and designed to meet the difference between the heat pump output and peak heat demand load.
Backflow	Rainwater harvesting Greywater reuse	Movement of fluid contrary to its intended direction of flow within an installation.
Backflow prevention arrangement	Rainwater harvesting Greywater reuse	An arrangement or device which is intended to prevent the contamination of water by backflow.
Back-up water supply	Rainwater harvesting Greywater reuse	A supply of potable (or wholesome) water from the public mains water supply or other potable source, that can supplement the non-potable supply in times of drought and/or heavy demand.
Battery bank	Solar photovoltaic Micro-wind turbine Micro-hydropower	A battery storage arrangement that stores surplus generated direct current electricity for future use.
Biological disinfection (greywater reuse)	Greywater reuse	Disinfection using aerobic or anaerobic bacteria to digest any unwanted organic matter in stored greywater.
Borehole	Heat pump	A narrow shaft bored in the ground to enable the installation of a heat pump collector circuit.
Buffer tank	Heat pump	A tank that contains a volume of water to absorb any extra heat generated by the heat pump in low load conditions which the building does not yet require.
Carbon dioxide (CO_2)	All technologies	A greenhouse gas that is believed to cause climate change and is emitted in a number of ways such as through the carbon cycle and through human activities such as the burning of fossil fuels.
Carbon neutral technology	Biomass	A technology that achieves net zero carbon emissions by balancing the volume of carbon dioxide released through combustion with the volume of carbon dioxide consumed during the growth and harvesting of the fuel.
Class 5 risk	Rainwater harvesting Greywater reuse	A risk of serious health hazard because of the possible concentration of pathogenic organisms, radioactive or very toxic substances, including: a) faecal material or other human waste b) butchery or other animal waste c) pathogens from any other source.
Clean Air Act	Biomass	The Clean Air Act 1993. The act provides the current legislative control relating to smoke emissions and emissions of other pollutants such as particles, sulphur dioxide, PAH and PCDD/F (dioxins and furans) which may be present in smoke.
Climate change	All technologies	The long-term alteration in global weather patterns, especially increases in temperature and storm activity, regarded as a potential consequence of the greenhouse effect.

Term	System context	Meaning
Code for Sustainable Homes	All technologies	The national standard for the sustainable design and construction of new homes. The Code aims to reduce our carbon emissions and create homes that are more sustainable. It applies in England, Wales and Northern Ireland.
Coefficient of Performance (CoP)	Heat pumps	The measure of efficiency of a heat pump unit expressed as a ratio or as a per cent of the output of a heat pump in heating mode.
Co-generation technologies	Micro-combined heat and power	A technology that can simultaneously generate both electricity and useful heat. Co-generation technologies can be heat-led (producing electricity as a secondary output to heat) or electricity-led (producing heat as a secondary output to electricity).
Collector circuit	Heat pumps	A circuit used to extract heat from ground or water sources to provide the renewable heat energy input to a heat pump unit.
Combination boiler	Solar thermal hot water	A boiler with the capability to provide domestic hot water directly, in some cases containing an internal hot water store.
Conservation area	All technologies	Local authorities have the power (under Section 69 of the Planning (Listed Buildings and Conservation Areas) Act 1990) to designate as conservation areas, any area of special architectural or historic interest. This means the planning authority has extra powers to control works and demolition of buildings to protect or improve the character or appearance of the area.
Cross-connection	Rainwater harvesting Greywater reuse	A physical hydraulic link (permanent or temporary) between a pipework system carrying non-wholesome water connected to a pipework system carrying mains wholesome water.
Defrost cycle	Heat pumps	A cycle where an air source heat pump operates in reverse to remove ice that has formed on the heat pump external evaporator coil as a result of the freezing during normal operation of moisture that is present in the air at low ambient temperatures.
Electricity distribution grid	Solar photovoltaic Micro-wind turbine Micro-hydropower	A combination of the national transmission network and regional distribution networks used to distribute mains electricity.
Electron	Solar photovoltaic	A subatomic particle which carries a negative electrical charge.
Energy performance certificate	All heat producing and electricity producing technologies	A certificate that provides a rating of the energy and carbon emission efficiency of a building using a grade from 'A' to 'G'. An 'A' rating is the most efficient, while 'G' is the least efficient.
Existing final circuit	Micro-combined heat and power	A final ring circuit feeding 13 Amp socket outlets.
External combustion engine	Micro-combined heat and power	An engine where an (internal) working fluid is heated by combustion of an external source, through the engine wall or a heat exchanger.
Fabric energy efficiency	All heat producing technologies	The efficiency of the built fabric (walls, floors, windows, roof etc.) of a building in relation to the energy demand for space heating and cooling expressed in kilowatt-hours of energy demand per square metre per year ($kWh/m^2/year$).
Fan coil unit	Heat pumps	A device consisting of a heating or cooling coil and a fan.
Feed-In Tariffs (FiTs)	Solar photovoltaic Micro-wind turbine Micro-hydropower Micro-combined heat and power	FiTs is a Government scheme that offers payments for surplus electricity generated through environmental technologies.
Finned heat exchanger	All heat producing technologies	A tubular heat exchanger that included fins to increase the efficiency of the heat transfer process.
Fish pass arrangement	Micro-hydropower	A physical arrangement that allows the safe passage of migrating fish such as salmon and sea trout.
Fish screen	Micro-hydropower	A physical arrangement that prevents the entry of fish into a micro-hydropower system.
Fossil fuels	Micro-combined heat and power	Fuels with a finite supply, formed by natural processes over thousands and millions of years, i.e coal, gas and oil.
Fuel cell	Micro-combined heat and power	An electrochemical cell that converts chemical energy from a fuel into electric energy.
Fuel poverty	All heat producing technologies	Occurs when, in order to heat its home to an adequate standard of warmth, a household needs to spend more than 10% of its income on total fuel use.

Term	System context	Meaning
Geothermal energy	Heat pumps	Energy generated and stored in the Earth. The Earth's geothermal energy originates from the formation of the planet, from radioactive decay of minerals, from volcanic activity, and from solar energy absorbed at the surface.
Geothermal water source	Heat pumps	A below ground water source that is heated by the earth's energy.
Glycol	Solar thermal hot water Heat pumps	A liquid which can be added to water in a closed loop pipework circuit such as a solar thermal hot water system or heat pump collector circuit to prevent freezing. The most common types are Ethylene Glycol and Propylene Glycol.
Green Deal	All technologies	The Coalition Government's flagship policy for improving the energy efficiency of buildings in Great Britain. The Green Deal will create a new financing mechanism to allow a range of improvement measures to reduce energy demand and improve energy efficiency measures to be installed in people's homes and businesses at no upfront cost.
Greenhouse gas	All heat producing and electricity producing technologies	Gases such as carbon dioxide that act as a shield that traps heat in the earth's atmosphere resulting in the greenhouse effect.
Greywater	Greywater reuse	Domestic wastewater excluding faecal matter and urine.
Heat exchanger	All heat producing technologies	A component in which heat is exchanged such as a coil within a hot water cylinder or a multiple plate heat exchanger such as a condenser in a heat pump unit.
Heat sink circuit	Heat pumps	A circuit connected to a heat pump unit in which the heat that is provided from the heat pump is distributed.
Immersion heater	Solar thermal hot water	An electrical water heating element.
Instantaneous centralised hot water system	Solar thermal hot water	A system with multiple domestic hot water outlets in which domestic hot water is heated instantaneously, on demand in a centralised location before being distributed to the domestic hot water outlets.
Internal combustion engine	Micro-combined heat and power	An engine in which the combustion of a fuel (normally a fossil fuel) occurs with an oxidizer (usually air) in a combustion chamber.
Low carbon technology	Solar thermal Heat pumps	Technologies considered to be low carbon in operation (powered at least in part by fossil fuels) and technologies that are zero carbon in operation (powered by 100% renewable energy).
Main channel	Micro-hydropower	Any shallow body of a river or a stream.
Microgeneration Certification Scheme (MCS)	All heat producing and electricity producing technologies	A scheme introduced by the Government to provide consumers with an assurance that microgeneration products and installation companies meet a robust set of standards.
Organic rankine cycle engine	Micro-combined heat and power	An engine that converts heat into work. The heat is supplied externally to a closed loop, which usually uses water. The organic rankine cycle engine is a type of steam-operated heat engine most commonly found in power generation plants.
Orientation	Solar thermal hot water Solar photovoltaic	The siting of a building or an elevation of a building in relation to the available solar energy.
Overshading	Solar thermal hot water Solar photovoltaic	Shading of solar thermal collectors or solar photovoltaic panels caused by obstructions such as buildings and trees. The degree of overshading may vary during different times of the day.
Permitted development rights	All technologies	Rights granted under The Town and Country Planning (General Permitted Development) Order 1995 (and subsequent amendments) to allow certain types of minor changes to your buildings without needing to apply for planning permission.
Photons	Solar photovoltaic	A discrete bundle (or quantum) of electromagnetic (or light) energy that is contained in solar energy. Photons within solar energy are used within solar photovoltaic systems.
Photovoltaic	Solar photovoltaic	The production of electric current at the junction of two substances exposed to light.
Physical disinfection	Rainwater harvesting Greywater reuse	Disinfection using filters to remove unwater matter in rainwater and greywater.
Point of use hot water system	Solar thermal hot water	A system in which domestic hot water for a single hot water outlet is heated at the point at which the outlet is located.

Term	System context	Meaning
Rainwater	Rainwater harvesting	Water arising from atmospheric precipitation.
Regular boiler	Solar thermal hot water	A boiler which does not have the capability to provide domestic hot water directly (i.e. not a combination boiler). It may nevertheless provide domestic hot water indirectly via a separate hot water storage cylinder.
Renewable energy	All heat producing and electricity producing technologies	Energy derived from natural processes that are replenished constantly, including electricity and heat generated from solar, wind, ocean, hydropower, biomass, geothermal resources, and biofuels and hydrogen derived from renewable resources.
Renewable Heat Incentive	All heat producing technologies	A proposed Government scheme, due to be introduced in late 2011 to provide financial support to consumers who install eligible renewable heat technology systems. Consumers will receive financial support in the form of pence per kilowatt hour of renewable heat energy generated.
Semiconducting materials	Solar photovoltaic	A semiconducting material is a material that behaves in between a conductor and an insulator. At ambient temperature, it conducts electricity more easily than an insulator, but less readily than a conductor. At very low temperatures, pure or intrinsic semiconductors behave like insulators. At higher temperatures or under light, intrinsic semiconductors can become conductive. The addition of impurities to a pure semiconductor can also increase its conductivity.
Shadow flicker	Micro-wind turbine systems	Flicker that occurs when the sun passes behind the hub of a wind turbine and casts a shadow over neighbouring properties.
Shunt pump	Solar thermal hot water	A circulation pump fitted to hot water service/plant to overcome the temperature stratification of the stored water.
Sluice	Micro-hydropower	A water channel that is controlled at its head by a gate. For example, a millrace is a sluice that channels water toward a water mill.
Solar irradiation	Solar thermal hot water / Solar photovoltaic	The density of the sun's energy reaching the earth. Under ideal conditions the density is approximately 1 kW/m^2 in Central Europe. The irradiation density on the surface of the earth depends on the location. With seasonal changes, times of overcast sky, and night time this amounts to an average solar irradiation of about 1000 kWh per m^2 and year.
Thermal stress	Solar photovoltaic	Stress that may occur due to overheating of a solar photovoltaic cell or cells with the potential to cause permanent damage to the cell(s).
Thermosiphon	Solar thermal hot water	The process where heated water expands becomes lighter and rises, and as water cools it contracts in volume, becomes heavier and falls.
Type AA air gap	Rainwater harvesting / Greywater reuse	A non-mechanical arrangement that provides a physical air gap between the spillover level of a cistern or water system connection and a back-up supply that is connected to the cistern or water system allowing for unrestricted discharge of the flow of non-wholesome water in order to prevent backflow into the wholesome water supply.
Type AB air gap	Rainwater harvesting / Greywater reuse	A non-mechanical arrangement used within a storage cistern that includes the provision of a weir overflow to introduce a physical air gap between the weir and the lowest point of a back-up supply that is connected to the cistern allowing for unrestricted discharge of the flow of non-wholesome water in order to prevent backflow into the wholesome water supply.
UV disinfection	Rainwater harvesting / Greywater reuse	A non-chemical approach to disinfection that uses UV-C light to penetrate a microorganism's outer cell membrane, pass through its cell body and disrupt its DNA preventing reproduction.
Watercourse	Micro-hydropower	Any flowing body of water including rivers, streams and brooks.
Weir	Micro-hydropower	A small overflow dam used to alter the flow characteristics of a river or stream.
Wholesome water	Rainwater harvesting / Greywater reuse	Water which is fit to use for drinking, cooking, food preparation or washing without any potential danger to human health by meeting regulatory requirements.
World Heritage site	All technologies	A World Heritage site is a cultural or natural site, monument, city or geographical habitat that is deemed irreplaceable and threatened, and thus deserving of protection and preservation by the United Nations Educational, Scientific and Cultural Organization (UNESCO). There are currently 28 World Heritage sites in the UK.
Zero carbon technology	Solar photovoltaic / Micro-wind turbine / Micro-hydropower	A technology that is powered 100% by renewable energy.

Acknowledgements

Photo credits

p.9 *L* © Econergy, *TR* © Kensa Heat Pumps, *BR* © Sussex Solar Ltd; **p.11** *T* © Sussex Solar Ltd, *B* © Sussex Solar Ltd; **p.27** © Econergy; **p.28** *L* © Leuthi Enterprises Limited, *TR* © Solar UK, *BR* © MannPower Consulting Limited; **p.30** *T* © Solar UK, *B* © Solar UK; **p.31** *TL* © Solar UK, *BL* © GreenGate Publishing Services, *TR* © Landis + Gyr; **p.40** © MannPower Consulting Limited; **p.45** *L* © ACO Technologies plc, *R* © Hansgrohe Group

The National Skills Academy for Environmental Technologies would like to thank the many people who contributed to this resource.

The National Skills Academy
ENVIRONMENTAL TECHNOLOGIES

Notes

The National Skills Academy
ENVIRONMENTAL TECHNOLOGIES